Introduction to Animal Anatomy

Introduction to Animal Anatomy

Matt Fuller

STATES
ACADEMIC PRESS
www.statesacademicpress.com

States Academic Press,
109 South 5th Street,
Brooklyn, NY 11249, USA

Visit us on the World Wide Web at:
www.statesacademicpress.com

ISBN: 978-1-63989-301-0 (Hardback)

Cataloging-in-Publication Data

Introduction to animal anatomy / Matt Fuller.
p. cm.
Includes bibliographical references and index.
ISBN 978-1-63989-301-0
1. Anatomy. 2. Biology. 3. Physiology. 4. Veterinary anatomy.
5. Veterinary physiology. I. Fuller, Matt.

QL805 .I58 2022
596.04--dc23

Table of Contents

Preface

The branch of biology which focuses on studying the structural organization of living organisms is known as anatomy. It is an inter-disciplinary field which encompasses the principles of developmental biology, evolutionary biology, embryology and phylogeny. Animal anatomy refers to the study of the members of the animal kingdom. It is categorized into vertebrate anatomy and invertebrate anatomy. The field of vertebrate anatomy focuses on the body plans of vertebrates and is further classified into fish anatomy, amphibian anatomy, reptile anatomy, bird anatomy and mammal anatomy. Invertebrate anatomy deals with simple invertebrate organisms. These range from eukaryotic organisms like paramecium to multicellular animals like dragonfly, octopus and lobsters. Study of animal tissues such as muscle tissues, nervous tissues and connective tissues also falls under this discipline. The topics included in this book on animal anatomy are of utmost significance and bound to provide incredible insights to readers. Some of the diverse topics covered herein address the varied branches that fall under this category. Coherent flow of topics, student-friendly language and extensive use of examples make this book an invaluable source of knowledge.

To facilitate a deeper understanding of the contents of this book a short introduction of every chapter is written below:

Chapter 1- The science that deals with the form and structure of animals is known as animal anatomy. All animals are complex multicellular organisms. They can be divided into two separate groups; vertebrates and invertebrates, each with a distinctive anatomy. This is an introductory chapter which touches on the various anatomies of both vertebrates and invertebrates.

Chapter 2- Animal cells are eukaryotic cells with a membrane-bound nucleus, lacking a cell wall. They contain all the necessary structures and constituents required for the survival of the organism. Tissues are a group of connected cells performing similar functions with different tissue groups varying in structure, function and origin. Animal tissues can be classified into various types which are examined in detail in this chapter.

Chapter 3- The animal body comprises of many organ systems that help in the everyday functioning of the organism. The muscular system, integumentary system, skeletal system, digestive system, respiratory system, excretory system, reproductive system, endocrine system, etc. are a few of the systems present in all animal bodies. In this chapter, each topic is discussed in detail for the benefit of the reader.

Chapter 4- The nervous system integrates and controls the various functions of the body, and detects and reacts to stimuli with the help of the sensory organs. The nervous system coordinates the actions and sensory information by transmitting signals to and from different parts of the body. The topics elaborated in this chapter will help in gaining a better perspective of the sensory organs and the nervous system in animals.

Chapter 5- The displacement of an animal from one place to another is called locomotion. Movement is the motion of body parts of the organism without changing the location of the organism. Locomotion helps the animal find food and shelter, escape from predators or even facilitate mating. This chapter covers the different types of locomotion in animals such as arboreal locomotion, terrestrial locomotion, aquatic locomotion, passive locomotion, etc. in a comprehensive manner.

Finally, I would like to thank the entire team involved in the inception of this book for their valuable time and contribution. This book would not have been possible without their efforts. I would also like to thank my friends and family for their constant support.

Matt Fuller

Chapter 1

Understanding Animal Anatomy

The science that deals with the form and structure of animals is known as animal anatomy. All animals are complex multicellular organisms. They can be divided into two separate groups; vertebrates and invertebrates, each with a distinctive anatomy. This is an introductory chapter which touches on the various anatomies of both vertebrates and invertebrates.

The term anatomy refers to the science that deals with the form and structure of animals. All animals are made up of cells, some of which are specialized to carry out different functions. Simple animals, such as sponges, are made up of only a few types of cell. In more complex animals, cells are organized into tissues, such as muscles and nerves that are necessary for movement. Tissues can form organs, such as the heart, which is used to pump blood around the circulatory system. All living organisms, with the exception of basic single-celled organisms, are made up of various different parts which enable them to function. The majority of an animal's body is made up of water, cells and a protective or supportive structure such as a skeleton made out of bone or a hard outer shell to protect it.

Body Systems

Animals have a number of different body systems, each of which does a specific job in order to keep the animal functioning. The heaviest body systems are the skeletal system (although it is not present in all animals), and the muscular system which connects the muscles to the brain so that they know what to do. Other body systems include the circulatory system, where blood is pumped around the body of the animal by its heart; the digestive system which breaks down food and extracts the nutrients from it so they can be turned into energy; and the respiration system which controls an animal's breathing. All body systems are vital in the working of a healthy animal.

Skeletons

Many animals have some kind of hard structure either on the inside of their body (known as an endoskeleton) for support or on the outside of their body (known as an exoskeleton) for both support and protection. Other animals may not have a hard support system at all, like jellyfish for example, which are held together by a thin layer of cells known as a membrane. Skeletons help to keep the animal stable by supporting its body shape, and some land animals are so large that their bodies even contain extra bony plates as the skeleton on its own is not strong enough to support the animal's weight.

Organs

Similar cells within the body of an animal form together in groups which are organized into tissues and various tissues make up different organs in an animal's body. There are a number of different organs found in the body of an animal, each of which has a specific task. Organs are linked to form organ systems, which carry out the vital processes in an animal's body that are key to it's survival, like the heart pumping the blood around the body or the skin of an animal, protecting it's delicate insides from the outside world.

Respiration

Animals need oxygen in order to survive and many animals get their oxygen by breathing the surrounding air into their lungs, which is passed through the body and exhaled as carbon dioxide (so animals breathe in oxygen and breathe out carbon dioxide). Respiratory systems differ between animal species, so birds and mammals for example process the oxygen in different ways. Some animals however, do not have lungs and instead use other methods in which they can get oxygen into their bodies. Fish live in the water and so use gills in order to extract oxygen from the water surrounding them. Gills are made up of thin, feathery flaps which draw the blood close to the surface of the skin allowing the oxygen particles to pass into it. Other animals, like some frog species also breathe through pores in their skin.

Senses

Typically, most animal species have five senses which are sight, smell, taste, touch and balance. Certain species of animal have more heightened senses than others, like a mole that lives under-ground so can barely see, to an eagle that can spot prey on the ground from miles up in the air. Some animal species have also developed a "sixth sense" such as echolocation in dolphins, whales and bats.

Movement

Being able to move independently is one of the factors that make an organism an animal. Movement is caused by the contracting of muscles within an animal's body, and is displayed in countless ways throughout the animal kingdom. Humans walk, cheetahs can run as fast as a car, ducks and geese waddle, fish swim, birds fly and some insects get around by jumping. Some adult animals however, such as barnacles and sponges become sessile by permanently fixing themselves to another animal or object, they are still animals however as they still respire and are able to move in their immature form.

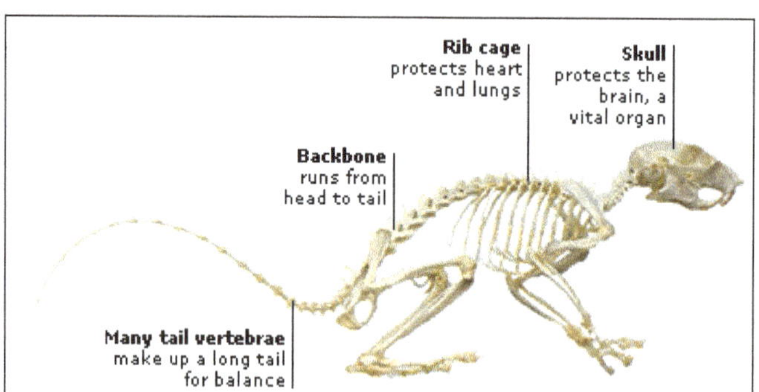

The study of the structure of living things is called anatomy. All animals are made up of CELLS, some of which are specialized to carry out different functions. Simple animals, such as sponges, are made up of only a few types of cell. In more complex animals, cells are organized into tissues, such as muscles and nerves that are necessary for movement. Tissues can form organs, such as the heart, which is used to pump blood around the CIRCULATORY SYSTEM. All animals with backbones have an internal framework of support, called an endoskeleton. Bony skeletons, such as that of the squirrel, are light to aid movement. When an animal is young, bones in the skeleton can grow in length. Some bones protect vital organs, while limb bones provide anchorage for muscles.

All animals need oxygen to survive. Simple animals exchange gases over the surface of their bodies. Insects, such as caterpillars, have openings along their bodies, called spiracles, through which air passes. Animals with lungs, such as birds and mammals, actively breathe.

Perfect Symmetry

Most animals are bilaterally symmetrical. If a penguin were cut in half from head to toe, the two halves would be mirror images of each other. Other animals, such as sea anemones, are radially symmetrical. They have no head or tail and can be cut into identical halves along many lines. Of the two types, animals that are bilaterally symmetrical tend to move more quickly and precisely.

Shark Anatomy

Like all fish, sharks have a backbone, breathe through gills, maneuvers using fins, and are ectothermic (cold-blooded). A shark's anatomy also bears the hallmarks of a predatory fish. They have a streamlined, torpedo-shaped body that allows them to cut easily through water to chase prey. They also have powerful jaws and sharp teeth.

Exoskeleton

Like all arthropods, lobsters have a hard outer casing, called an exoskeleton, made up of plates formed from a substance called chitin. The plates meet at flexible points, such as the leg joints. This exoskeleton provides anchorage for muscles and protection from predators. It also provides support for movement on land and prevents excess water loss.

Cells

Animal cells are typically just 0.02 mm (1/1,250 in) across. Although they can be extremely varied, they share common features. Cells are surrounded by a skin called a membrane and contain a jelly-like fluid called cytoplasm. All the processes needed for life, such as producing energy from food, removing waste, and growth takes place inside cells.

Cell Components

Inside an animal cell, the cytoplasm contains structures called organelles that have a variety of functions, from storing vital substances to destroying bacteria. The most important organelle is called the nucleus, which carries genetic information, controlling how the cell behaves. Another organelle, the mitochondrion, produces energy from food.

Vertebrate Anatomy

Vertebrates are members of the subphylum Vertebrata, under the phylum Chordata and under the kingdom Animalia. Animals that possess bilateral symmetry can be divided into two groups, protostomes and deuterostomes, based on their patterns of embryonic development. The deuterostomes, whose name translates as "second mouth," consist of two phyla: Chordata and Echinodermata. Echinoderms are invertebrate marine animals that have pentaradial symmetry and a spiny body covering; the phylum includes sea stars, sea urchins, and sea cucumbers. The phylum Chordata contains two groups of invertebrate chordates, but the most conspicuous and familiar members of Chordata are the vertebrates.

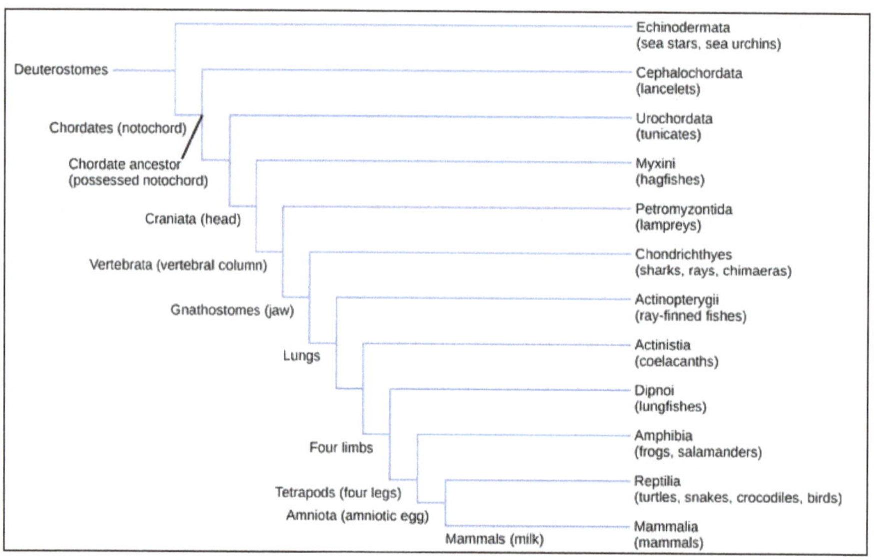

Phylum chordate all chordates are deuterostomes, possessing a notochord.
Vertebrates are differentiated by having a vertebral column.

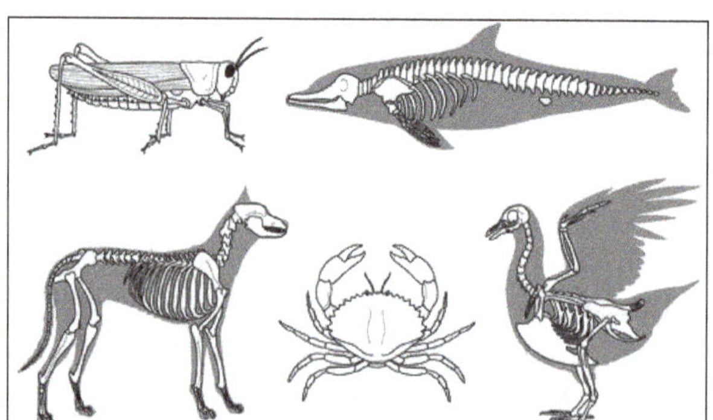

As chordates, all vertebrates have a similar anatomy and morphology with the same qualifying characteristics: a notochord, a dorsal hollow nerve cord, pharyngeal slits, and a post-anal tail.

However, the subphylum Vertebrata is distinguished from the phylum Chordata by the development of the notochord into a bony backbone. Vertebrates include the amphibians, reptiles, mammals, and birds, as well as the jawless fishes, bony fishes, sharks, and rays.

More than 64,000 species of vertebrates have been described, but the extant vertebrate species represent only a small portion of all the vertebrates that have existed. Vertebrates range in size from the frog species *Paedophryne amauensis* (as small as 7.7 mm (0.3 inch)) to the blue whale (as large as 33 m (110 ft)). Vertebrates comprise about 4 percent of all described animal species; the remainders are invertebrates, which lack backbones.

Anatomy and Morphology

All vertebrates are built along the basic chordate body plan: a stiff rod running through the length of the animal (vertebral column), with a hollow tube of nervous tissue (the spinal cord) above it and the gastrointestinal tract below. In all vertebrates, there is a mouth at anterior end of the animal and an anus before the posterior end of the body. There is a tail posterior to the anus during at least one phase of the animal's development.

Vertebrate Classification

Vertebrates are the largest group of chordates, with more than 62,000 living species. Vertebrates are grouped based on anatomical and physiological traits. The traditional groups include Agnatha, Chondrichthyes, Osteichthyes, Amphibia, Reptilia, Aves, and Mammalia.

Animals that possess jaws are known as gnathostomes, meaning "jawed mouth." Gnathostomes include fishes and tetrapods (amphibians, reptiles, birds, and mammals). Tetrapod's can be further divided into two groups: amphibians and amniotes. Amniotes are animals whose eggs are adapted for terrestrial living; this group includes mammals, reptiles, and birds. Amniotic embryos, developing in either an externally-shelled egg or an egg carried by the female, are provided with a water-retaining environment and are protected by amniotic membranes.

Vertebrate Structure and Function

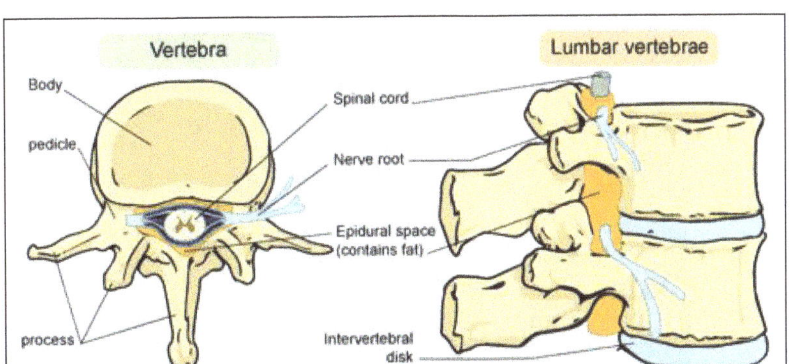

Human Vertebral Column and Vertebrae: The human vertebral column consists of 33 vertebrae. Two enlarged vertebrae are shown here.

The vertebral column, or backbone, is made up of a series of repeating, or metameric, structures called vertebrae (singular, vertebra) that extend from the head to the tail along the dorsal (top) side of the body. The function of the vertebral column is to provide the central support structure of the endoskeleton, allowing an organism to hold its shape. It also serves to house and protect the spinal cord that that lies within the column. A portion of a vertebral column containing the spinal cord is shown in Figure below.

The structure of individual vertebrae as well as the number of vertebrae that make up the backbone varies between different species. Humans, for example, have a backbone consisting of 33 vertebrae (including several that are fused together), whereas snakes can have as many as 500! The general structure of a vertebra, as shown in Figure below, consists of two main regions: a centrum and a neural arch. The centrum is a disk-shaped, solid portion that provides strength to the backbone. The neural arch is located on the dorsal side of the centrum and forms the hole through which the spinal cord passes. Some species may have spines that extend either from the centrum or the neural arches of the vertebrae.

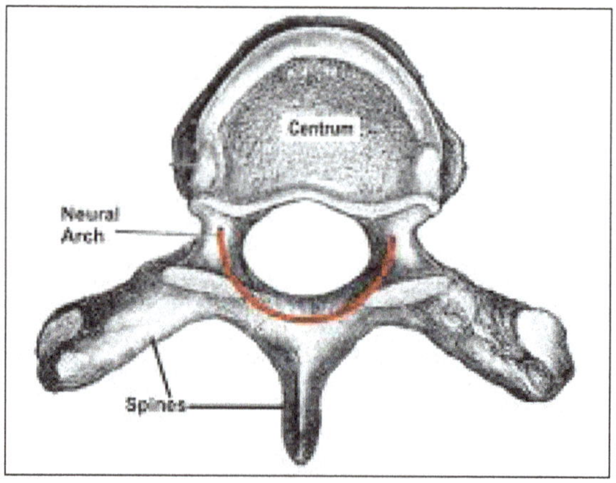

The vertebra is composed of two main regions: The centrum and the neural arch. The centrum is a solid, disc-shaped section that provides most of the strength of the backbone. The neural arch encloses the spinal column and is located dorsal (on the back side) to the centrum. Different species of vertebrates may have a number of processes called spines that extend from the main structure of the vertebra. In some species the spines can be quite pronounced.

Vertebrates are characterized by having a pronounced head, like the one of this Antarctic toothfish that houses the brain and several sensory organs including eyes. Centralization of the nervous system in a distinct head region is called cephalization.

All vertebrates are highly cephalized, meaning that their nervous system is centralized with a brain located in a distinct head; the head region also contains major sensory organs such as a pair of complex eyes (cephalization). Although many invertebrates exhibit a certain level of cephalization with nervous tissue concentrated in the anterior, or "head," end of the animal, the head region of vertebrates is more distinct than that of many invertebrates. Compare the fish species shown in Figure below with an invertebrate chordate, the lancelet, from the class Cephalochordata. When looking at a lancelet, it is difficult to distinguish the head from the tail because there is no distinct head region, although there is a brain located specifically at one end of the animal's body.

Within this well-defined head region, the vertebrate brain is enclosed in a cranium, or skull, from which the vertebral column extends. The vertebral column (along with the skull) makes up the core of the endoskeleton. A typical vertebrate endoskeleton, like the ones shown in Figure below, usually includes the following components:

- A vertebral column.
- A cranium.
- Limbs/appendages (usually in pairs).
- Limb girdles (hips and shoulders in humans).
- Visceral arches (located in the neck region and form either gills or jaws and middle ear components).

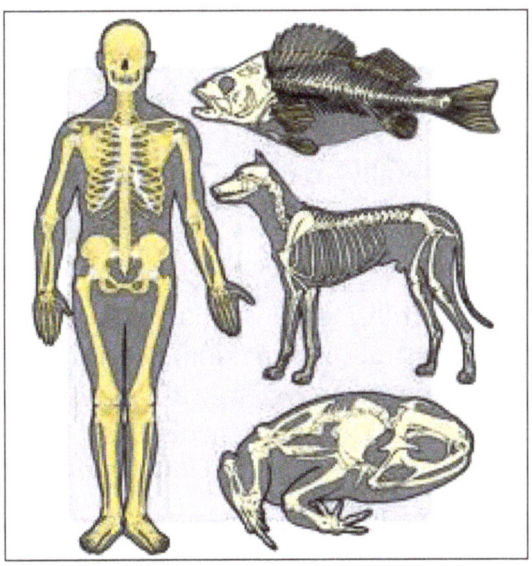

Vertebrate Endoskeletons: The vertebrate endoskeleton includes a vertebral column, cranium, limbs, and limb girdles.

The vertebrate endoskeleton may be composed entirely of cartilage or composed of both cartilage and bone. An exclusively cartilaginous endoskeleton is found in several fish classes, but the majority of vertebrates have endoskeletons made up primarily of bone tissue. What is the difference between cartilage and bone? Cartilage is less dense and therefore lighter and more flexible than bone. As a result, cartilage does not have the strength to support the large bodies of many of the higher vertebrate species. Bone tissue consists of a dense extracellular (outside the cell) matrix containing a protein called collagen. Collagen is a fibrillar protein, forming long fibers that form the basis of the bone matrix and contribute to the strength of the tissue. This matrix is also highly mineralized, meaning that it contains high levels of inorganic elements such as calcium, magnesium, and phosphate.

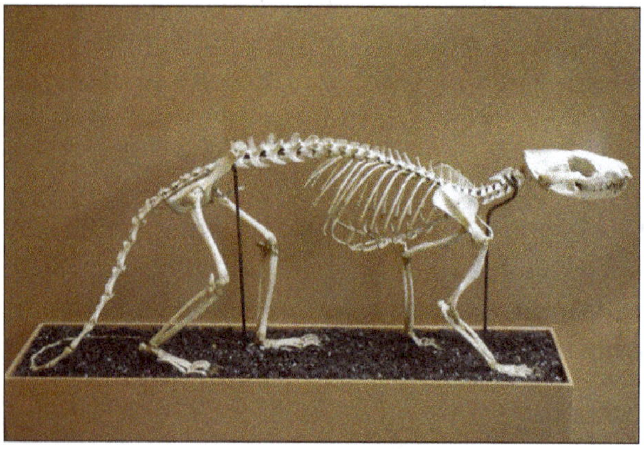

The endoskeleton of a tiger quoll: A typical vertebrate skeleton, like this one, includes a cranium, a vertebral column, limbs, and limb girdles where the limbs attach to the main part of the body.

Bone cells called osteocytes are distributed throughout this matrix. Osteocytes are derived from osteoblasts, and it is the osteoblasts that are responsible for secreting the collagen and minerals that make up the bone tissue. In the embryos of bony vertebrates, the endoskeleton is initially formed from cartilage that is converted to bone during development. Bone tissue is also associated with bone marrow in many vertebrates. Bone marrow gives rise to many of the blood cells of the circulatory system.

The endoskeleton of all but two vertebrate classes includes jaws equipped with teeth that are used for chewing food and, in some cases, for attacking prey. Most vertebrates also have at least one pair of limbs that are generally located laterally, or on the sides of the body.

Organ Systems

In addition to the skeletal system, vertebrates have many other complex organ systems including a closed circulatory system with a ventral (bottom side) heart having two to four chambers, a respiratory system, a well-developed muscular system with a muscular mouth and pharynx, an excretory system, and an immune system. Most of these organ systems are located within a well-developed coelom, or body cavity.

Like chordates in general, the muscular system of vertebrates is made up of repeating units called myomeres. Muscles attached to the endoskeleton coordinate movements by alternating between contracted (shortened) and relaxed (extended) states. Most regions of the endoskeleton have muscles attached on opposite sides so that movements are accomplished by the reciprocal contraction and relaxation of oppositely positioned muscles. The majority of vertebrate species have adaptive immune systems. An adaptive immune system is one that can respond to infections by generating new forms of molecules, called antibodies that are specifically designed to target the particular invading organism.

Probably the most striking feature of vertebrates is their complex nervous system. The vertebrate nervous system consists of two main components: the central nervous system (brain and spinal cord), which is shown in Figure below, and the peripheral nervous system, which basically consists of any part of the nervous system that is not the brain or spinal cord.

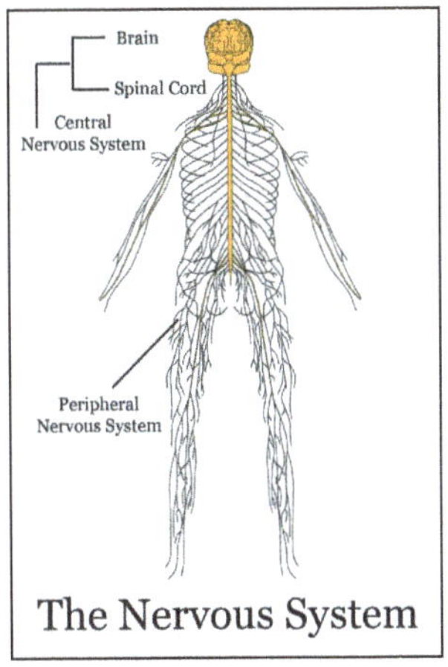

The Human Nervous System: The vertebrate nervous system includes a brain and spinal cord. It also includes a body-wide network of nerves, called peripheral nerves. They connect the spinal cord with the rest of the body.

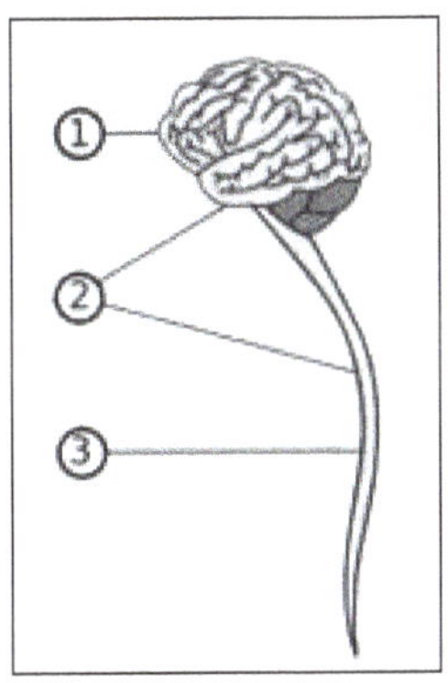

A diagram showing the components of the central nervous system. The central nervous system (2) consists of a brain (1) located in the anterior, or head, region of an organism and a long spinal cord (3) that extends from the brain to the tail region, or posterior end, of an organism. The central nervous system is housed within the cranium (brain) and the vertebral column (spinal cord) of the endoskeleton.

Nerve fibers that extend to the muscular system and all other organs within the body are part of the peripheral nervous system. Both of these systems are run by nerve cells, or neurons, that are interconnected to create a vast network of cells throughout the body. Neurons communicate with each other to pass information between tissues of the body and regions of the brain through electrochemical signals. These signals are either electrical impulses that can be transmitted between cells or molecules secreted by one nerve cell that can bind to an adjacent nerve cell to relay information. The level of organization of nervous systems varies greatly within the vertebrate subphyla, from a relatively simple system in some primitive fish species to the enormously complex nervous system of primates.

Reproduction

Vertebrates reproduce sexually, and almost all species have separate male and female sexes. Fertilization can be either external or internal depending on the species. In general, aquatic vertebrates undergo external fertilization, whereas terrestrial vertebrates carry out internal fertilization, although there are exceptions to this rule.

Development also varies from species to species. The majority of vertebrate species lay eggs, and their embryos develop externally. This is called oviparity. There is also a significant number of species, including some fish species that retain their eggs within the female body until they hatch and emerge from the mother, a feature called ovoviviparity. A more advanced version of internal development, called viviparity, is seen in the majority of mammal species. In these cases, not only does the mother retain eggs for internal development, but there are also special organs that develop to transfer nutrients from the mother's body to the developing offspring.

Neural Crest Cells

Another critical feature of vertebrates is the neural crest. The neural crest forms at the edge of the developing nerve cord in vertebrates, as shown in Figure below. Neural crest cells are extremely important in vertebrate development because soon after they are formed, they migrate to other regions of the body where they either become or cause other cells to become critical components of a variety of different tissues including cells of the nervous, muscular, skeletal, and hormone-producing endocrine tissues. Defects in the development or migration of neural crest cells lead to a variety of congenital birth defects.

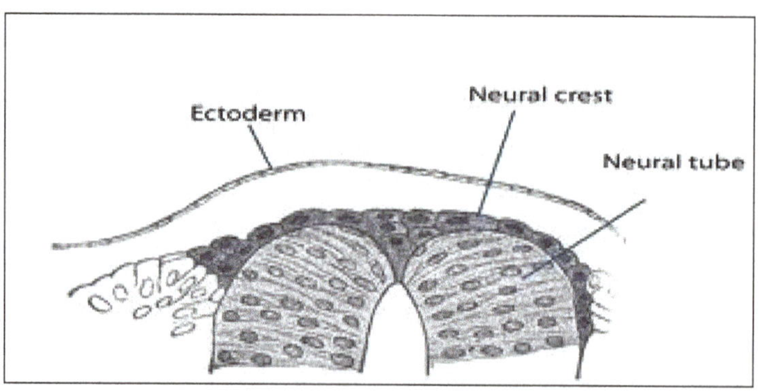

The vertebrate neural crest is a group of cells that form at the edge of the developing neural tube during vertebrate embryonic development. These cells will soon migrate to various regions of the body to become components of different tissues. The neural tube will go on to develop into the brain and spinal cord.

Fish Anatomy

Fish are animals that are cold-blooded, have fins and a backbone. Most fish have scales and breathe with gills. Approximately, 22,000 species of fish began evolving 480 million years ago. The large-mouth bass illustrated above has the typical torpedo-like (fusiform) shape associated with many fishes.

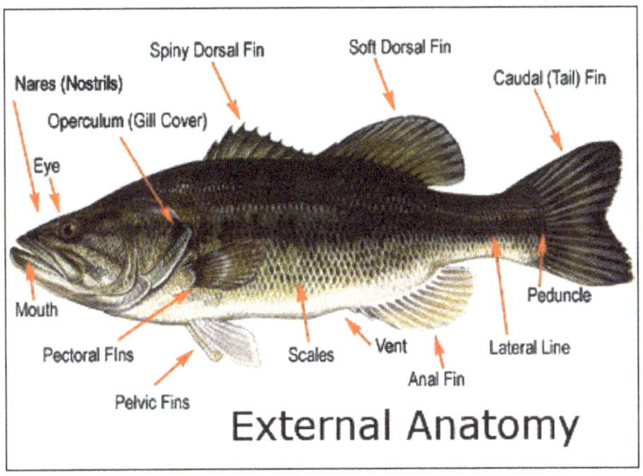

Fins are appendages used by the fish to maintain its position, move, steer and stop. They are either single fins along the centerline of the fish, such as the dorsal (back) fins, caudal (tail) fin and anal fin, or paired fins, which include the pectoral (chest) and pelvic (hip) fins. Fishes such as catfish have another fleshy lobe behind the dorsal fin, called an adipose (fat) fin. The dorsal and anal fins primarily help fish to not roll over onto their sides. The caudal fin is the main fin for propulsion to move the fish forward. The paired fins assist with steering, stopping and hovering.

Scales in most bony fishes (most freshwater fishes other than gar that have ganoid scales, and catfish which have no scales) are either ctenoid or cycloid. Ctenoid scales have jagged edges and cycloid ones have smooth rounded edges. Bass and most other fish with spines have ctenoid scales composed of connective tissue covered with calcium. Most fishes also have a very important mucus layer covering the body that helps prevent infection. Anglers should be careful not to rub this "slime" off when handling a fish that is to be released.

In many freshwater fishes the fins are supported by spines that are rigid and may be quite sharp thus playing a defensive role. Catfish have notably hard sharp fins, of which anglers should be wary. The soft dorsal and caudal fins are composed of rays, as are portions of other fins. Rays are less rigid and frequently branched.

The gills are the breathing apparatus of fish and are highly vascularized giving them their bright red cover. An operculum (gill cover) that is a flexible bony plate that protects the sensitive gills. Water is "inhaled" through the mouth, passes over the gills and is "exhaled" from beneath the operculum.

Fish can detect color. The eyes are rounder in fish than mammals because of the refractive index of water and focus is achieved by moving the lens in and out, not reshaping the lens as in mammals.

Paired nostrils, or nares, in fish are used to detect odors in water and can be quite sensitive. Eels and catfish have particularly well developed senses of smell.

The mouth's shape is a good clue to what fish eat. The larger it is, the bigger the prey it can consume.

Fish have a sense of taste and may sample items to taste them before swallowing if they are not obvious prey items. Most freshwater fishes in Florida are omnivorous (eating both plant and animal matter). Some are primarily piscivorous (eating mostly other fish). The imported grass carp is one of the few large fishes that are primarily herbivorous (eating plants). Fish may or may not have teeth depending on the species. Fish like chain pickerel and gar have obvious canine-shaped teeth. Other fish have less obvious teeth, such as the cardiform teeth in catfish which feel like a roughened area at the front of the mouth, or vomerine teeth that are tiny patches of teeth, for example, in the roof of a striped bass' mouth. Grass carp and other minnows have pharyngeal teeth modified from their gill arches for grinding that are located in the throat.

The lateral line is a sensory organ consisting of fluid filled sacs with hair-like sensory apparatus that are open to the water through a series of pores (creating a line along the side of the fish). The lateral line primarily senses water currents and pressure, and movement in the water.

The vent is the external opening to digestive urinary and reproductive tracts. In most fish, it is immediately in front of the anal fin.

Internal Fish Anatomy

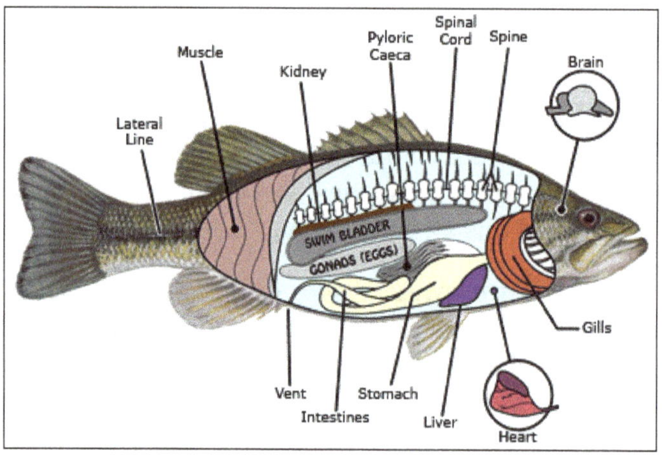

As different as a man may be from a fish, both creatures share some fascinating similarities in basic structure and function. And the closer one looks, the more complex life becomes. The smallest units of life are microscopic cells, and some organisms--such as an ameba--are no larger than a single cell. In larger multicellular creatures, individual cells that are similar in structure and perform a specific function are grouped into tissues, and tissues may be grouped into even more complex and specialized structures called organs. These organs perform the basic bodily functions such as respiration, digestion, and sensory reception. Man and fish share such organs as the brain, stomach, liver, and kidneys. Other organs appear in different forms in different organisms; for example, the lungs in humans and the gills in fish are very different but both provide the same basic function of respiration. Finally, some organs (such as the fish's swim bladder) are simply not present in man. A number of other vital organs, such as the spleen and pancreas, may also be present but are smaller and more difficult to locate. A largemouth bass destined for the frying pan makes an excellent specimen because this species is large enough for easy examination. For anglers brave enough to do some investigating while filleting their next fish, a fascinating learning experience awaits! Don't forget to examine the stomach content, as this can give clues to what the fish was feeding on, where in the water column and what lures to present. For instance, small shad probably came from open water and a shad-imitating lure may be your best bet; crayfish would suggest working a soft plastic along the bottom.

Spine

The primary structural framework upon which the fish's body is built; connects to the skull at the front of the fish and to the tail at the rear. The spine is made up of numerous *vertebrae*, which are hollow and house and protect the delicate spinal cord.

Spinal Cord

Connects the brain to the rest of the body and relays sensory information from the body to the brain, as well as instructions from the brain to the rest of the body.

Brain

The control center of the fish, where both automatic functions (such as respiration) and higher behaviors occur.

Lateral Line

One of the fish's primary sense organs; detects underwater vibrations and is capable of determining the direction of their source.

Swim or Air Bladder

A hollow, gas-filled balance organ that allows a fish to conserve energy by maintaining neutral buoyancy (suspending) in water. Fish caught from very deep water sometimes need to have air released from their swim bladder before they can be released and return to deep water, due to the difference in atmospheric pressure at the water's surface. Species of fish that do not possess a swim bladder sink to the bottom if they stop swimming.

Gills

Allow a fish to breathe underwater. These are very delicate structures and should not be touched if the fish is to be released.

Kidney

Filters liquid waste materials from the blood; these wastes are then passed out of the body. The kidney is also extremely important in regulating water and salt concentrations within the fish's body, allowing certain fish species to exist in freshwater or saltwater, and in some cases (such as snook or tarpon) both.

Stomach and Intestines

Break down (digest) food and absorb nutrients. Fish such as bass that are *piscivorous* (eat other fish) have fairly short intestines because such food is easy to chemically break down and digest. Fish such as tilapia that are *herbivorous* (eat plants) require longer intestines because plant matter is usually tough and fibrous and more difficult to break down into usable components. A great deal about fish feeding habits can be determined by examining stomach contents.

Pyloric Caeca

This organ with fingerlike projections is located near the junction of the stomach and the intestines. Its function is not entirely understood, but it is known to secrete enzymes that aid in digestion, may function to absorb digested food, or do both.

Vent

The site of waste elimination from the fish's body. It is also the entry to the genital tract where eggs or sperm are released.

Liver

This important organ has a number of functions. It assists in digestion by secreting enzymes that break down fats, and also serves as a storage area for fats and carbohydrates. The liver also is important in the destruction of old blood cells and in maintaining proper blood chemistry, as well as playing a role in nitrogen (waste) excretion.

Heart

Circulates blood throughout the body. Oxygen and digested nutrients are delivered to the cells of various organs through the blood, and the blood transports waste products from the cells to the kidneys and liver for elimination.

Gonads (Reproductive Organs)

In adult female bass, the bright orange mass of eggs is unmistakable during the spawning season, but is still usually identifiable at other times of the year. The male organs, which produce milt for fertilizing the eggs, are much smaller and white but found in the same general location. The eggs (or *roe*) of certain fish are considered a delicacy, as in the case of caviar from sturgeon.

Muscles

Provide movement and locomotion. This is the part of the fish that is usually eaten, and composes the fillet of the fish.

Amphibian Anatomy

Amphibians are comprised of the orders Anura (frogs and toads), Caudata (including salamanders, newts and sirens) and Gymnophiona (caecilians). In Australia, we are primarily concerned with the health of anurans.

Amphibians undergo a metamorphosis that is unique among vertebrates; from a juvenile, obligatory aquatic form with gills (tadpoles) to a semi-aquatic adult form with lungs. Adult amphibians also have the unique ability to breathe transdermally (through the skin), meaning changes to the skin through disease, trauma or mishandling can have significant health impacts.

Early recognition and a deep understanding of the significant diseases affecting amphibians will be crucial to halting the current decline in species biodiversity seen in Australia. Although not of great commercial interest in Australia, native amphibians act as sentinels of change in aquatic environments. The anatomical descriptions in Figures are not exhaustive but are intended to help field guide users communicate the location and extent of gross external anatomical changes.

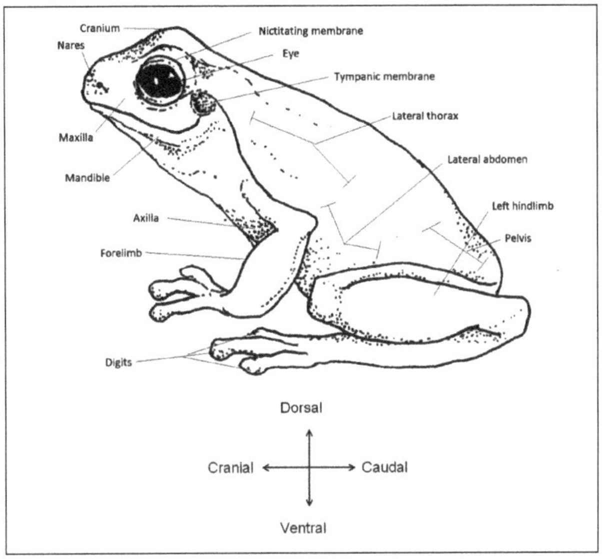

External anatomy of frog—left lateral view.

Aquatic animal diseases significant to Australia: Identification field guide 5th edition.

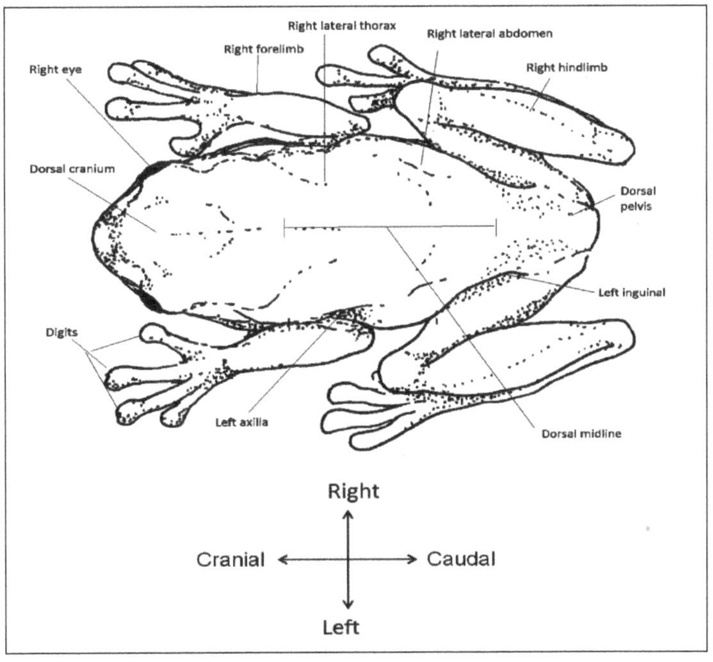

External anatomy of frog—dorsal view.

Reptiles Anatomy

Most reptiles are ectotherms (cold-blooded): They maintain their body temperature by absorbing heat from the environment, so their body temperature changes with environmental temperature. In contrast, most mammals and birds are endotherms (warm-blooded): They are able to maintain a constant body temperature despite changes in the environmental temperature. Snakes, lizards, and chelonians have a 3-chambered heart. Crocodilians have a 4-chambered heart. Eggs are fertilized internally (inside the body). Some reptile species lay eggs from which the offspring hatch, and others give birth to live young.

Determining the sex of most reptiles (particularly snakes) can be tricky and in most cases is best left to a veterinarian, a herpetologist, or an experienced breeder or dealer. The males and females of some species of lizards have distinguishing characteristics such as size, color, or scale pattern. However, some reptiles are hard to reliably identify as male or female by visible characteristics. Male turtles have a longer tail than females, and the cloacal opening in males is closer to the tail tip. Among semiaquatic reptiles, males are smaller and have longer claws. They might also have a spur on the hindlegs. Terrestrial turtles and tortoises have distinct differences in the shape of their underbelly (plastron): it is concave (rounded) in males and flat in females. Some male tortoises also have a larger pair of gular scales on the front end of the underbelly.

Snakes

Snakes range in size from the size of a worm to many feet long. Their skin contains scales that may be smooth or ridged. The scales on the belly (called scutes) are thicker than those on the sides and back of a snake to provide protection as it moves.

Snakes have some clear differences from other reptiles. They have no limbs, no moveable eyelids, and no ear openings. The teeth of most nonvenomous snakes form 2 rows in the upper jaw and 1 row in the lower jaw. The teeth are curved backward to help keep struggling prey from escaping. Venomous snakes have grooved or hollow fangs that they use to inject venom into their prey. Snakes' skulls contain joints to allow the snake to swallow large prey.

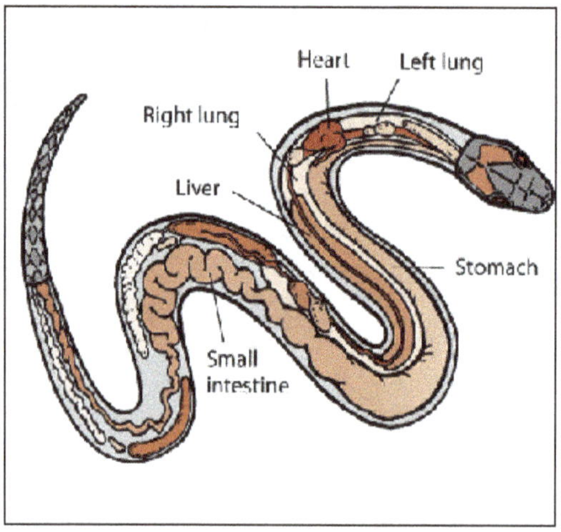

Internally, snakes have many sets of ribs to support their length. These ribs are somewhat flexible to allow the prey to move through the snake's body. In addition, the organs inside a snake are long and narrow to fit inside the body cavity. Snakes move in different ways, including the familiar undulating crawl (slithering), side-winding, and an accordion-like movement used to climb trees.

Snakes rely primarily on their senses of smell and touch. Their forked tongues bring small air particles into the mouth, where an organ in the roof of the mouth identifies odors. Although snakes do not have external ears, they have an ear bone that can detect vibrations of sound waves.

Lizards

There are more than 4000 species of lizards, ranging in size from a few inches to the Komodo dragon, the largest lizard, which can reach up to 10 feet in length. Lizards that are commonly kept as pets include geckos, anoles, iguanas, skinks, chameleons, and agamids (including bearded dragons).

Most lizards have dry skin made up of scales. The scales of lizards vary from the smoother scales of skinks to rough scales or even spikes. In many species, the tail is fragile and can break easily. It can regenerate, although the new growth may look different.

Lizards are adapted to many different environments. Some are good swimmers, and others spend most of their time in trees. Many have clawed feet that help them climb and cling.

Like snakes, lizards use their tongues to help them smell. The tongue captures particles of air and brings them into the mouth, where a specialized organ can detect various smells. Lizards have external ears and appear to be able to hear well than snakes. Most lizards have eyelids that clean and protect their eyes when they blink. A few, however, have fixed eyecaps like snakes.

Some lizards have developed special features to help them survive. Chameleons and some other species such as anoles can change the color of their scales to blend in with their surroundings. The males of some species have a loose flap of skin called the dewlap that can be extended to either intimidate a predator or to help attract a mate. And, as mentioned above, a lizard's tail can break off, which can help it escape from predators.

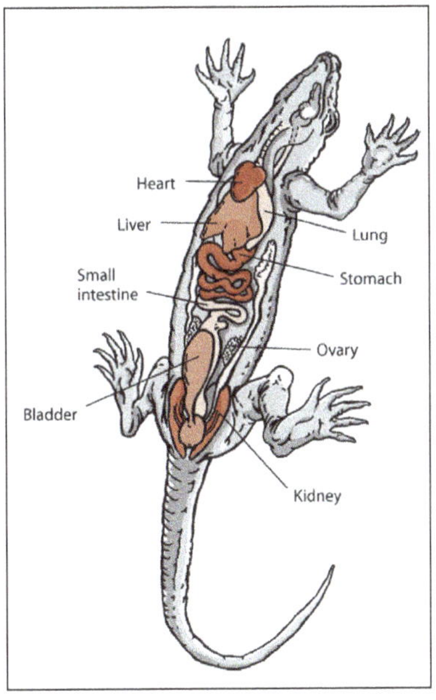

Turtles and Tortoises

Turtles and tortoises belong to a group of reptiles known as chelonians. They are easily distinguished by their hard shells that protect their upper and lower bodies. The upper covering is known as the carapace, and the bottom portion is called the plastron. The words "turtle" and "tortoise" are often used interchangeably, and in different parts of the world they can mean different things. The word "turtle" can refer to any chelonian or specifically to chelonian species that live mostly in water (including sea turtles and those found in ponds or rivers). Tortoises live only on land and do not swim.

Chelonians range in size from small (shells 3 to 4 inches in length) to very large (shell length of up to 8 feet). Many chelonians can be quite long-lived. Some species of tortoises have been known to survive for up to 150 years, and some aquatic turtles may live for 70 years.

The shells of turtles and tortoises are made up of a large number of bones that are covered by large scales called scutes. The shell is part of a turtle's skeleton and is permanently attached to the turtle at the spine. Some turtles can tuck their head, legs, and tails inside their shells, but others cannot. The shell enlarges as the turtle grows, either by replacing old scutes with larger ones during shedding or by enlarging the diameter of existing scutes. In some cases, this can help determine the age of the animal.

Like snakes, turtles and tortoises hear by feeling vibrations in the ground or water. Many have good eyesight and a good sense of smell to help them locate food. They do not have teeth; rather, their mouths have a sharp edge that they use to bite or tear food. Many turtles and tortoises are omnivores, meaning they eat both plants and animals (such as insects or worms), although some eat only plants.

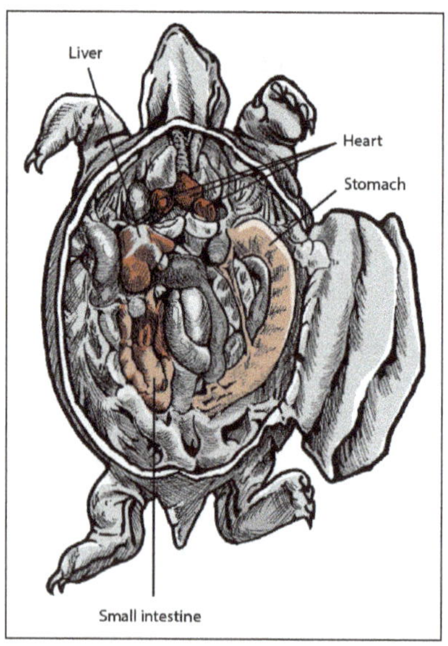

Anatomy of a turtle.

Bird Anatomy

Birds are physiologically different from any other creature on the face of the earth. While they need to eat, drink, and breathe the same as we do, the parts of their bodies that carry out these functions differ drastically from our own.

External Bird Anatomy

- Beak: A bird's beak serves many purposes--such as eating, grooming, and of course, singing! The beak is an extension of the bird's jaw bone and is covered in keratin, the same substance that makes up our fingernails. The top part of the beak is called the cere and is where the bird's nostrils, or nares, are located.
- Eye: Whoever came up with the phrase "eagle eye" wasn't joking--birds have an extraordinarily accurate vision. The eye holds scores of receptor cells, called rods and cones that translate whatever the bird looks at to the image that it sees. To give an idea of how sharp their sight is, humans typically have around 200,000 of these cells per millimeter inside of their eyes. Some birds, particularly birds of prey, have five times that many.
- Wings: A bird's wings are constructed of a series of small thin bones similar to miniature versions of the bones in human arms. Externally, the wings are home to several different kinds of feathers: the Primary Flight Feathers, the Secondaries, the Main and Lesser Coverts, the Tertials, and the Alula.
- Foot: The feet and legs of birds vary greatly depending on the species. Generally, the legs, feet, and claws are structured to allow a bird to take off, land, climb, and grasp with them. Since birds spend most of their lives perching, the feet and legs are covered with tougher skin than the skin on the rest of the bird's body.
- Tail: During flight, a bird's tail acts much like the tail of an airplane--it's used as a rudder to help the bird steer. The muscles of the tail also aid in helping the bird expand its lungs to take in extra air when needed.
- Anus: The anus is the external opening through which the bird passes its waste.

Internal Anatomy

Birds are just as different from us on the inside as they are on the outside:

- Brain: Being called a bird brain isn't necessarily a bad thing--in fact, some may take it as a compliment! Birds are extremely intelligent creatures, and as any bird owner knows, they never fail to surprise us with their capacity for learning.
- Spinal Column: Like all vertebrates, birds have a spinal column that runs the length of their bodies, and encases the delicate spinal cord. The spinal cord is part of the central nervous system and, in essence, acts as the brain's "messenger." When the bird decides that he wants to move, the spinal cord relays the message from the brain to the muscles that correspond to the desired body part, causing movement.
- Trachea: The trachea is a long tube that runs from the bird's throat to its lungs, and transports fresh air for the bird to breathe.
- Esophagus: The bird's esophagus is a narrow tube that transports food from the mouth to the crop, where it will be stored until it is digested.
- Lung: Much like human lungs, avian lungs serve to diffuse air throughout the bird's bloodstream. They are unique, however, in the fact that they have small air sacs that

allow air to flow through the lung in only one direction, ensuring a constant supply of fresh oxygen.

- Crop: In the same way that a chipmunk stores food in its cheeks, birds store food in their crops. The crop is composed of layers of muscle tissue and holds and softens the food until it's ready to be passed on to the gizzard.
- Gizzard: A gizzard is a structure composed of tough muscle tissue which contains roughage that is used to grind the bird's food into a pulp. When the food is sufficiently ground, it is passed into the bird's intestine.
- Kidney: Liquids that the bird ingests are passed into the kidneys, which filter out any waste to be expelled from the bird later.
- Heart: Much like our human hearts, a bird's heart is divided into four chambers and serves to pump oxygen-rich blood throughout the body. Because birds are such high-energy animals, their hearts beat much faster than those of mammals. Some bird species have a resting heart rate of over 500 beats per minute.
- Liver: A bird's liver acts much like a large filter, and rids the bird of any toxins in its body.
- Ureter: The ureter is a tube that extends from the kidney to the cloaca, and allows liquid waste to be expelled from the bird's body.
- Intestines: A bird's intestines work to digest the food that is pumped into them from the gizzard, absorbing the nutrients that the bird needs to function. After the food is digested, the waste is pushed into the rectum.
- Rectum: The rectum allows waste to be expelled from the bird's body.

Mammals Anatomy

Mammals are vertebrates that possess hair and mammary glands. Several other characteristics are distinctive to mammals, including certain features of the jaw, skeleton, integument, and internal anatomy. Modern mammals belong to three clades: monotremes, marsupials, and eutherians (or placental mammals).

Characteristics of Mammals

The presence of hair is one of the most obvious signs of a mammal. Although it is not very extensive on certain species, such as whales, hair has many important functions for mammals. Mammals are endothermic, and hair provides insulation to retain heat generated by metabolic work. Hair traps a layer of air close to the body, retaining heat. Along with insulation, hair can serve as a sensory mechanism via specialized hairs called vibrissae, better known as whiskers. These attach to nerves that transmit information about sensation, which is particularly useful to nocturnal or burrowing mammals. Hair can also provide protective coloration or be part of social signaling, such as when an animal's hair stands "on end."

Mammalian integument, or skin, includes secretory glands with various functions. Sebaceous glands produce a lipid mixture called sebum that is secreted onto the hair and skin for water resistance and lubrication. Sebaceous glands are located over most of the body. Eccrine glands produce sweat, or perspiration, which is mainly composed of water. In most mammals, eccrine glands are limited to certain areas of the body, and some mammals do not possess them at all. However, in primates, especially humans, sweat figures prominently in thermoregulation, regulating the body through evaporative cooling. Sweat glands are located over most of the body surface in primates. Apocrine glands, or scent glands, secrete substances that are used for chemical communication, such as in skunks. Mammary glands produce milk that is used to feed newborns. While male monotremes and eutherians possess mammary glands, male marsupials do not. Mammary glands likely are modified sebaceous or eccrine glands, but their evolutionary origin is not entirely clear.

The skeletal system of mammals possesses many unique features. The lower jaw of mammals consists of only one bone, the dentary. The jaws of other vertebrates are composed of more than one bone. In mammals, the dentary bone joins the skull at the squamosal bone, while in other vertebrates; the quadrate bone of the jaw joins with the articular bone of the skull. These bones are present in mammals, but they have been modified to function in hearing and form bones in the middle ear. Other vertebrates possess only one middle ear bone, the stapes. Mammals have three: the malleus, incus, and stapes. The malleus originated from the articular bone, whereas the incus originated from the quadrate bone. This arrangement of jaw and ear bones aids in distinguishing fossil mammals from fossils of other synapsids.

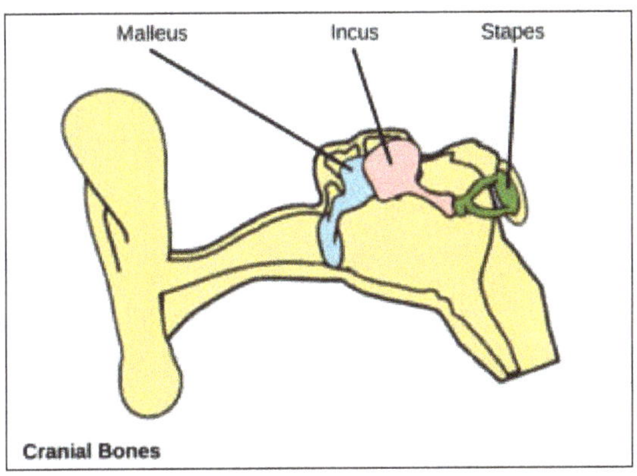

Bones of the mammalian inner ear are modified from bones of the jaw and skull.

The adductor muscle that closes the jaw is composed of two muscles in mammals: the temporalis and the masseter. These allow side-to-side movement of the jaw, making chewing possible, which is unique to mammals. Most mammals have heterodont teeth, meaning that they have different types and shapes of teeth rather than just one type and shape of tooth. Most mammals are diphyodonts, meaning that they have two sets of teeth in their lifetime: deciduous or "baby" teeth, and permanent teeth. Other vertebrates are polyphyodonts, that is, their teeth are replaced throughout their entire life.

Mammals, like birds, possess a four-chambered heart. Mammals also have a specialized group of cardiac fibers located in the walls of their right atrium called the sinoatrial node, or pacemaker, which determines the rate at which the heart beats. Mammalian erythrocytes (red blood cells) do not have nuclei, whereas the erythrocytes of other vertebrates are nucleated.

The kidneys of mammals have a portion of the nephron called the loop of Henle or nephritic loop, which allows mammals to produce urine with a high concentration of solutes, higher than that of the blood. Mammals lack a renal portal system, which is a system of veins that moves blood from the hind or lower limbs and region of the tail to the kidneys. Renal portal systems are present in all other vertebrates except jawless fishes. A urinary bladder is present in all mammals.

Mammalian brains have certain characteristics that differ from other vertebrates. In some, but not all mammals, the cerebral cortex, the outermost part of the cerebrum, is highly folded, allowing for a greater surface area than is possible with a smooth cortex. The optic lobes, located in the midbrain, are divided into two parts in mammals, whereas other vertebrates possess a single, undivided lobe. Eutherian mammals also possess a specialized structure that links the two cerebral hemispheres, called the corpus callosum.

Invertebrate Anatomy

Invertebrate is any animal that lacks a vertebral column, or backbone, in contrast to the cartilaginous or bony vertebrates. More than 90 percent of all living animal species are invertebrates. Worldwide in distribution, they include animals as diverse as sea stars, sea urchins, earthworms, sponges, jellyfish, lobsters, crabs, insects, spiders, snails, clams, and squid. Invertebrates are especially important as agricultural pests, parasites, or agents for the transmission of parasitic infections to humans and other vertebrates. Invertebrates serve as food for humans and are key elements in food chains that support birds, fish, and many other vertebrate species.

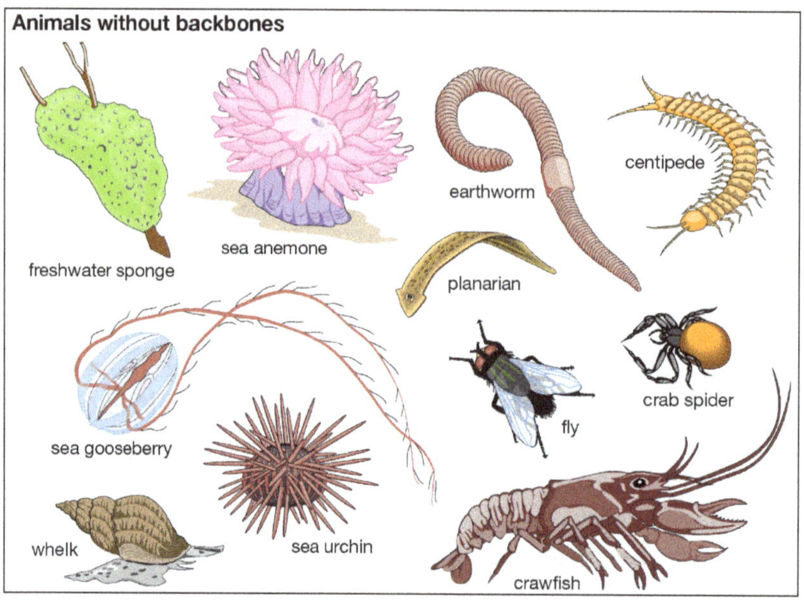

Invertebrates Selected invertebrate animals.

Apart from the absence of a vertebral column, invertebrates have little in common. Indeed, they are distributed between more than 35 phyla. In contrast, all vertebrates are contained within a single phylum, the Chordata. (Phylum Chordata also includes the sea squirts and some other invertebrate groups.) Invertebrates are generally soft-bodied animals that lack a rigid internal skeleton for the attachment of muscles but often possess a hard outer skeleton (as in most mollusks, crustaceans, and insects) that serves, as well, for body protection.

Arthropod Anatomy

Arthropods are not only the largest phylum of invertebrates. They are by far the largest phylum of the animal kingdom. Roughly 80 percent of all animal species living on Earth today are arthropods.

There are more than a million known species of arthropods. There may actually be ten times that many. Arthropods include insects, spiders, lobsters, and centipedes. The arthropods pictured in Figure below give just a hint of the phylum's diversity.

Arthropod Diversity: Dust mites are among the smallest of arthropods. Japanese spider crabs are the largest.

Structure and Function of Arthropods

Arthropods range in length from about 1 millimeter to 4 meters (about 13 feet). They have a segmented body with a hard exoskeleton. They also have jointed appendages. The body segments are the head, thorax, and abdomen. In some arthropods, the head and thorax are joined together as a cephalothorax.

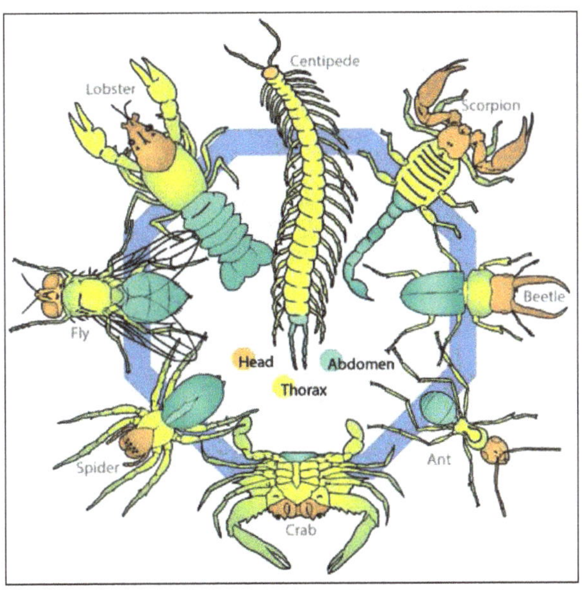

Arthropod Body Plan Notice the three body segments of each organism.

The arthropod exoskeleton consists of several layers of cuticle. The exoskeleton prevents water loss and gives support and protection. It also acts as a counterforce for the contraction of muscles. The exoskeleton doesn't grow as the animal grows. Therefore, it must be shed and replaced with a new one periodically through life. This is called molting.

The jointed appendages of arthropods may be used as legs for walking. Being jointed makes them more flexible. Try walking or climbing stairs without bending your knees and you'll see why joints are helpful. In most arthropods, the appendages on the head have been modified for other functions. Figure below shows some of head appendages found in arthropods. Sensory organs such as eyes are also found on the head.

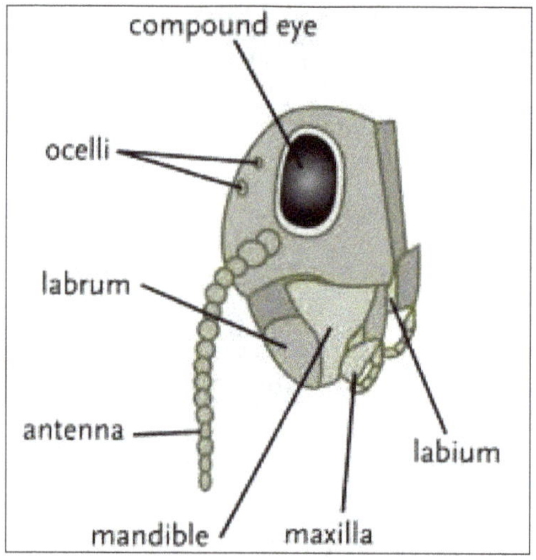

Arthropod Head: Arthropods have evolved a variety of specialized appendages and other structures on their head.

Some arthropods have special excretory structures. They are called coxal glands and Malpighian tubules. Coxal glands collect and concentrate liquid waste from blood. They excrete the waste from the body through a pore. Malphigian tubules carry waste from the digestive tract to the anus. The waste is excreted through the anus.

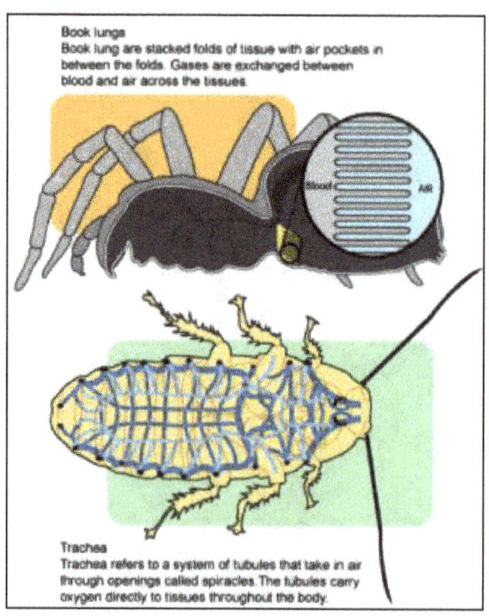

How Terrestrial Arthropods Breathe Air. Terrestrial arthropods have respiratory structures that let them breathe air.

Like mollusks and annelids, aquatic arthropods may have gills to exchange gases with the water. Terrestrial arthropods, on the other hand, have special respiratory structures to exchange gases with the air. These are described in figure above.

Underwater Spiders

In the ponds of northern Europe lives a tiny brown spider that spends its entire life underwater. But just like land spiders, it needs oxygen to breathe. So, how does this spider breathe? Does it use book lungs? No. In fact, aquatic spiders, known as "diving bell spiders," have gills. Every so often, the spider leaves its underwater web to visit the surface and bring back a bubble of air that sticks to its hairy abdomen. It deposits the bubble into a little silk air tank. This "diving bell" is a gill that sucks oxygen from the water, allowing the spider to stay underwater for up to 24 hours.

Chordate Anatomy

A chordate is an animal that belongs to the phylum Chordata, which is part of the Deuterostomes kingdom. Organisms in the Deuterostomes kingdom have a distinct characteristic: Their anus develops before their mouth in early embryonic stages. The phylum Chordata includes a wide range of organisms, as it is comprised of all vertebrates, which are organisms with a backbone, and many invertebrates; organisms that don't have a backbone. There are three subphyla to Chordata: Cepahlochordata, Urochordata, and Vertebrata. Chordates generally have bilaterally symmetric bodies, though a few exceptions exist, and they share distinct characteristics.

Characteristics of Chordates

Chordates have all of these characteristics, at least at some point in their lives. In many cases, these characteristics only appear during embryonic development.

Pharyngeal Slit

In invertebrates, this is used in feeding as a filter, while in vertebrate fish; it develops into gill arches, the function of which is to support gills. Human embryos have gillslits, but they disappear before we are born and the tissue develops into other structures in the head and neck. Generally, the tissue is modified for various purposes in terrestrial vertebrates, such as for the jaws and the Eustachian tube connecting the ear to the throat.

Dorsal Hollow Nerve Cord

This is a tube made of nerve fibers that develop into the central nervous system, consisting of the brain and spinal cord, in vertebrates. This nerve cord is protected by the vertebrae, which are the bones that make up our backbone. What makes chordates different is that, in non-chordates, the nerve cords are solid and are either ventral or lateral within the body.

Notochord

This flexible skeletal rod that's made of cartilage is what gives chordates their name. The notochord runs between an animal's digestive tube and its nerve cord and provides support for the body. It is the precursor of a backbone, or spine, in vertebrates.

Post-Anal Tail

This is a posterior elongation of the body that helps propel aquatic animals in water, provides balance, and is used by some terrestrial vertebrates to attract mates and signal when danger is near. This tail shrinks in humans and other apes into a tailbone during embryonic development.

Subphyla of Chordates

Subphylum Vertebrata

This is the subphylum that we, and all other vertebrates, belong in. Vertebrata is often referred to as Craniata because the organisms in this subphylum have a head with a protective cranium, which most of us call a skull. Vertebrates also have a brain encased in the skull, highly developed internal organs, a closed circulatory system, and unique sensory and motor cranial nerves. Fish, amphibians, reptiles, mammals, and birds are all part of his subphylum, and thanks to the presence of bone or cartilage, their fossils are easily found.

Subphylum Cephalochordata

This literally means "head cords". Cephalochordates are also called Lancelets, and they are filter-feeding marine animals with elongated, small, segmented, and soft bodies that make it difficult to find their fossils. They are usually found in soft bottoms as they bury themselves in the substrate, exposing their anterior part only (near the head) and using a row of tentacles to bring food into their mouth. Cephalochordates resemble fish, as we can see in the picture below, and swim like them, but they have no scales, backbone, limbs, or brain. They outdo fish, however, in the number of pharyngeal gill slits that they have.

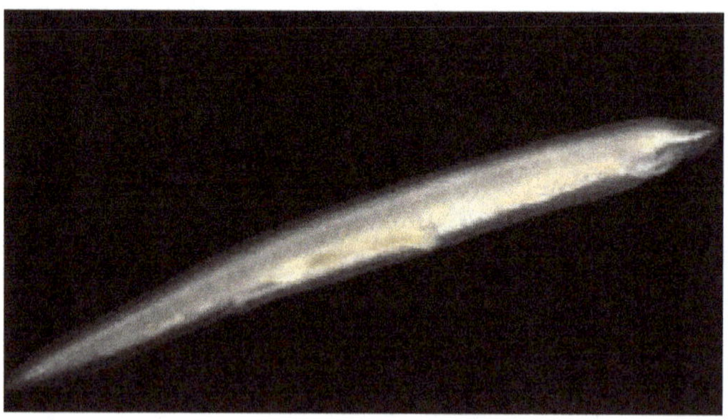

Branchiostoma Lanceolatum.

Subphylum Urochordata

Meaning "tail cords", these are also called Tunicates. This is another subphylum whose organisms' fossils are difficult to find, since the bodies have no hard parts. It includes sea squirts and sea salps. These are barrel-shaped, non-segmented filter-feeding marine animals. The larval stage usually has a tail and is free-swimming, but it eventually attaches to a hard substrate and loses the tail as it transforms into the adult form. By the time these organisms are adults, they only display one characteristic of chordates, and that is pharyngeal gill slits.

Examples of Chordates

Lampreys

The lamprey is a chordate that belongs to the subphylum vertebrata. It is a jawless fish that lives as a filter-feeder in its many larval years, then transforms into a parasitic adult that has an oral disk filled with teeth that it uses to latch onto other fish. Note, however, that some adults don't feed but rather live off reserves that they obtained as larvae. The adult has external gill slits for breathing, a cartilaginous skeleton, a nerve cord and a notochord. Larval lampreys use a mucus-secreting organ for trapping food particles in their pharynx, and have a post-anal tail. A lamprey larva is shown below.

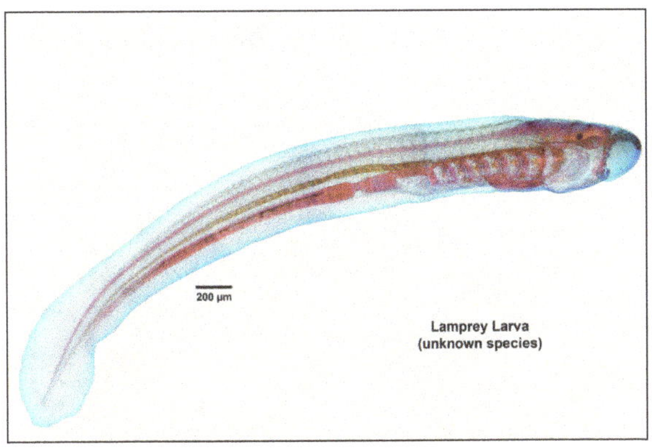

Lamprey larva unlabelled.

Sea Squirts

The sea squirt is a Urochordate, or Tunicate. As we can see in the picture below, its body is barrel-shaped and is attached to the substrate. The larvae, which are tadpole-like, possess a notochord, a dorsal nerve, pharyngeal slits, and a post-anal tail. The adults are filter-feeders, having two openings, called the siphons. The siphons are used in filter-feeding by drawing water into one side and straining the phytoplankton, which is the sea squirt's food source, then pumping out the filtered water through the other siphon.

Red spotted sea squirt.

These two examples are of relatively simple and small organisms, but remember that the phylum

Chordata comprises a wide variety of animals. It's amazing that such small animals can share a significant amount of characteristics with birds, mammals and plenty of other animals.

Mollusca Anatomy

Molluscs have a huge range of shapes and sizes even with a single group like the Gastropods. They, however, also have many features in common.

Shell

Many molluscs have shells. The edge of the mantle secretes a shell of calcium carbonate, chitin and conchiolin. Chitin is a tough, semitransparent carbohydrate. The exoskeletons of insects are made from chitin. Conchiolin is a complex protein which forms chambers which hold and bond to the crystals of calcium carbonate.

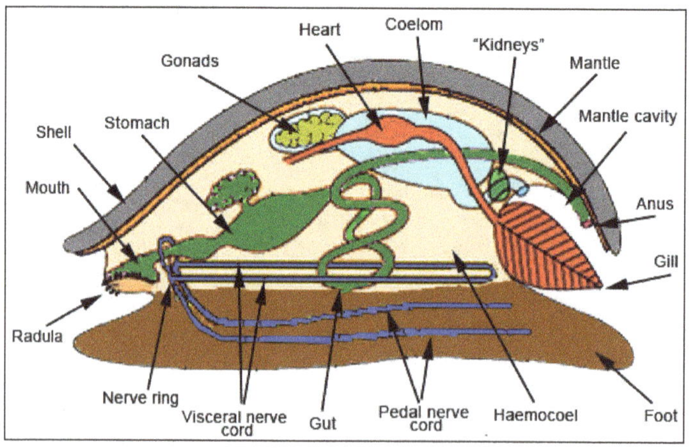

Anatomy of a "generalized mollusc".

Mantle and Mantle Cavity

Molluscs also have an extension of the body wall called the mantle. This fleshy tissue is responsible for secreting the shell. The mantle encloses the mantle cavity which contains the gills, anus and excretory pores.

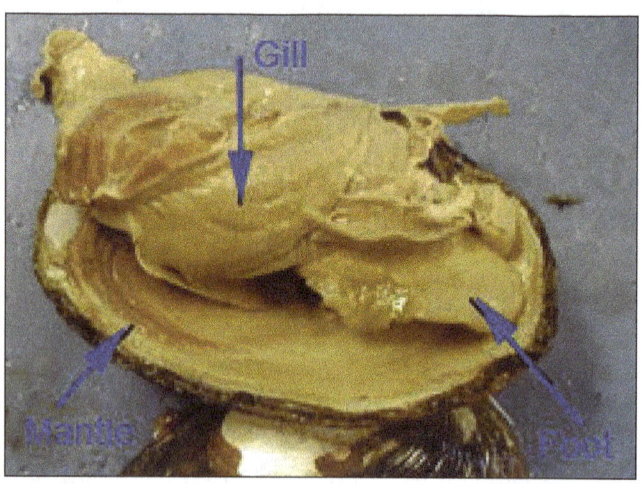

Gills

Most molluscs have one pair or only one feather-shaped gill found in the mantle cavity. Water enters near the bottom and leaves near the top. The gill filaments have cilia (like tiny hairs), one type of cilia beat to drive water through the mantle cavity, the other two types clean the gills. Each gill has an incoming blood vessel connected to the hemocoel and another leading to the heart.

Foot

The underside of most molluscs consists of a muscular foot, which has different purposes in different classes. For example,in gastropods the foot secretes mucuswhich acts as a lubricant to help movement; in many shells the foot acts as a sucker attaching the animal to a hard surface. In bivalves, the foot is used for burrowing into the sediment; in cephalopods it is used for jet propulsion, and the tentacles and arms are formed from the foot.

Radula

Most molluscs have a muscular mouths with a radula ("tongues") lined with many rows of teeth made from chitin. The teeth are replaced from the rear as they wear out. The radula is used to scrape bacteria and algae off rocks in herbivorous gastropods or to bore holes into other animals' shells or skeletons in carnivorous gastropods.

Mucus, formed in the mouth, mixes with the food and is pushed towards the stomach by cilia. Undigested materials are compressed and packaged, then released through the anus into the mantle cavity and are carried away from the animals in the water currents.

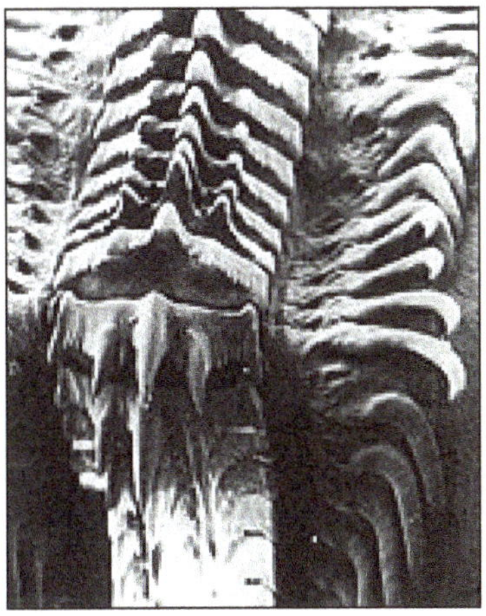

Coelom

The coelom is a fluid-filled cavity that develops within the body. It functions as a hydrostatic skeleton and provides a space in which the internal organs can be suspended.

A hydrostatic skeleton is a structure found in many cold-blooded organisms and soft-bodied animals consisting of a fluid-filled cavity, the coelom, surrounded by muscles. The pressure of the fluid and action of the surrounding muscles are used to change an organism's shape and produce movement, such as burrowing or swimming.

Haemocoel

A haemocoel is a series of spaces between the organs of organisms with open circulatory systems, like most arthropods and molluscs. Blood and other fluids circulate through the haemocoel. It encloses most of the internal organs.

In open circulatory systems (found in crustaceans, insects, molluscs and other invertebrates) the blood does not circulate entirely within vessels but is collected from the gills, pumped through the heart, and released directly into spaces (the haemocoel) in the tissues from which it returns to the gills and then to the heart. There are no capillaries to return blood to the heart.

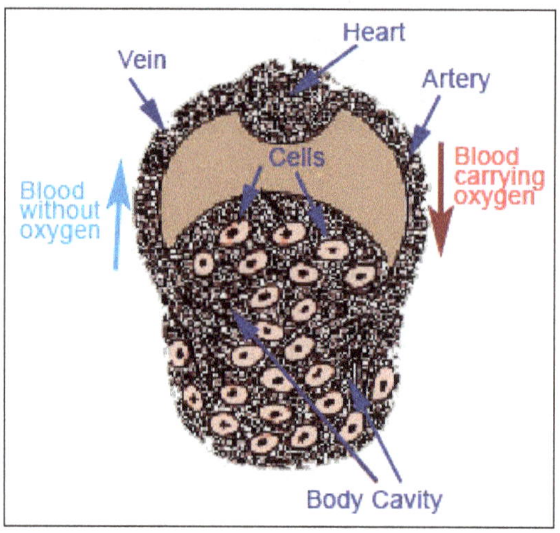

Heart

The heart is an important part of the circulatory system, pumping blood into the haemocoel, as well as an important part the reproductive and excretory systems.

Kidneys

Excretion is carried out by a pair of "kidneys" (nephridia), tubular structures that collect fluids from the coelom and exchange salts and other substances with body tissues as the fluid passes along the tubules for excretion. The kidneys empty into the mantle cavity. The heart and "kidneys" is an important part of the circulatory system as well as the reproductive and excretory systems. The "kidneys" are an important part of the excretory system as well as an important part the reproductive and circulatory systems.

Nervous System

Molluscs have two pairs of main nerves (three in bivalves) the visceral cords serving the internal organs and the pedal ones serving the foot. Only some species of molluscs have a brain.

Most molluscs have a head with eyes, and all have a pair of tentacles with sensors, also on the head, that detect chemicals, vibrations and touch. The eyes of the giant squid are the largest in the animal kingdom, almost the size of dinner plates.

Reproduction

The simplest molluscan reproductive system relies on external fertilization. Two gonads are located next to the coelom and release ova or sperm into it. The kidneys extract them from the coelom and release them into the mantle cavity from which they can reach the water or land. Some molluscs use internal fertilization and/or are hermaphrodites, functioning as both sexes; both of these methods require more complicated reproductive systems.

After fertilization, marine mollusc eggs grow into planktonic larvae which eat floating particles. Eventually the larva sinks to the seafloor and metamorphoses into the adult form.

Annelid Anatomy

Annelid is a phylum name Annelida, also called segmented worm, any member of a phylum of invertebrate animals that are characterized by the possession of a body cavity (or coelom), movable bristles (or setae), and a body divided into segments by transverse rings, or annulations, from which they take their name. The coelom is reduced in leeches, and setae are lacking a few specialized forms, including leeches. A major invertebrate phylum of the animal kingdom, the annelids number more than 9,000 species distributed among three classes: the marine worms (Polychaeta), which are divided into free-moving and sedentary, or tube-dwelling, forms; the earthworms (Oligochaeta); and the leeches (Hirudinea).

Structure and Function in Annelids

A polychaete commonly called the feather duster. These animals have elaborate feeding tentacles that emerge from the head and look like feathers.

The overall structure of annelids does not vary nearly as much as the structure of mollusks. All annelids have a worm-shaped, segmented body. One main variable in annelid anatomy is the number and organization of bristles and appendages protruding from the body. There is also a lot of variation between species in the number and form of feeding tentacles on the animal. Some of the parapodia and tentacles can be extremely elaborate and cause the animals to look quite unlike worms. One example of this is seen in feather dusters, a family of polychaetes shown in Figure below. However, beneath the feathery-looking protrusions of this animal, is a cylindrical worm-shaped body whose internal anatomy is very similar to that of a plain looking earthworm.

Overall Body Organization

Like flatworms and roundworms, annelids are bilaterally symmetrical triploblasts. Recall that bilateral symmetry means that they can only be divided into two equal halves by a cut along the middle of the anterior-posterior, or front-back, axis. Triploblasts develop from three embryonic cell layers: the ectoderm, which forms the outer surface of the animal and the nervous system, the mesoderm, which forms the middle layers of the organism, including muscle and circulatory tissues, and the endoderm, which forms the gut. Similar to mollusks, annelids are protostomes. Recall that protostomes exhibit spiral, determinant cleavage in their early embryonic cell divisions, and the blastopore ultimately forms the mouth. The body of an annelid is divided into three main regions: the prostomium, the trunk, and the pygidium. These are shown in Figure below.

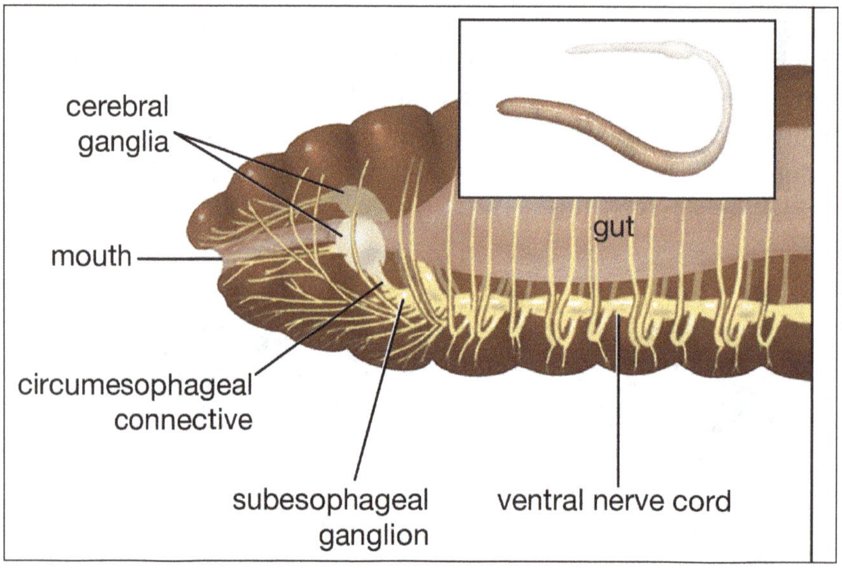

The three main regions of the annelid body.

The prostomium is the segment at the anterior, or front, end of the worm. The mouth usually lies just behind this segment. The region of the body surrounding the mouth is called the peristomium. Peristomium means "around the mouth," and this region makes up the back section of the head. The trunk includes the many segments that make up the bulk of the worm body and is located between the prostomium and the pygidium. The pygidium is the posterior, or back, end of the worm and houses the anus. Annelids grow by adding segments to the region just in front of, or anterior to, the pygidium. Segmented worms are also capable of regrowing regions of the worm that are broken off. This process is called regeneration, and the ability to regenerate depends on how much of the worm and what part of the body has been lost.

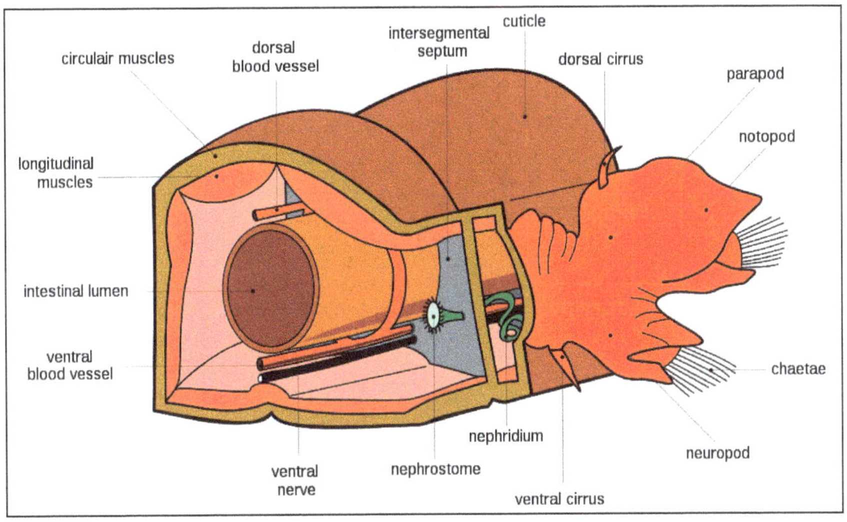

The coelom runs the length of the worm and is separated into segments by thin layers of tissue called septa (singular septum). Organs that span these compartments, such as the digestive tract and major blood vessels, pass through the septa. A cross section of an annelid body segment, showing the organization of various internal organs, is shown in Figure.

A cross-section of an annelid body (polychaete): This diagram shows the arrangement of organs within each segment. It also highlights the detailed architecture of the parapodium, which is divided into a number of regions: the parapod, notopod, neuropod, and several sensory projections called cirri (singular cirrus). The space between the intestine and the muscles of the body wall is the coelom, or body cavity.

Annelids commonly secrete a thick protective substance called the cuticle that covers the outer body wall and helps to keep the worm moist. The cuticle is made up of proteins and lipids.

Head Structure

There is a lot of variation in the head region among annelids. In general, species that have a sedentary or non-predatory, burrowing existence have simple heads that cannot be distinguished from their back ends unless you look very carefully. They do not usually have complex sensory organs because they really have no need for them. In contrast, active predatory annelids can have fairly complex head regions with eyes, tentacles, and sensory projections called palps and antennae. Predatory polychaetes also have jaws with teeth made up of a carbohydrate substance called chitin.

Exceptions to this general trend of simpler head structures for sedentary worms are found in a number of sedentary, marine polychaete species that have long, feathery feeding tentacles extending from their heads. While the worm's body usually remains burrowed or enclosed in a tube-shaped structure, these tentacles emerge from the burrow and spread out into the water to trap food particles.

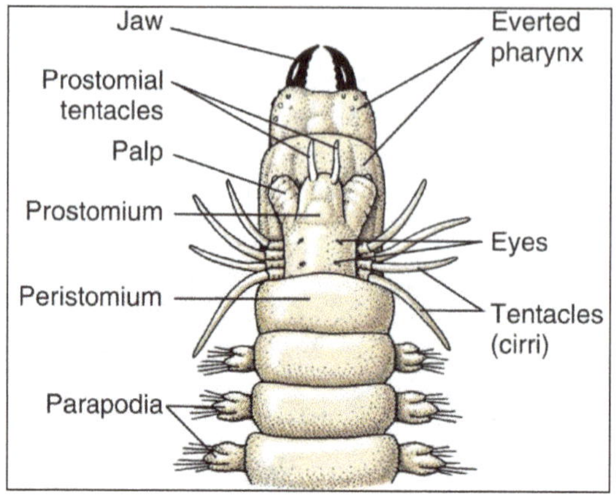

A diagram of the head of a predatory polychaete showing the antennae, palps, jaws, tentacles, and eyes.

Closed Circulatory System

Unlike most mollusks, annelids have a closed circulatory system. In a closed circulatory system, blood is always contained within blood vessels. The annelid circulatory system includes two major blood vessels that run the length of the worm: the dorsal blood vessel along the top and a ventral blood vessel along the bottom. The ventral vessel transports blood from the head to the tail region, and the dorsal vessel transports blood from the tail to the head. These two vessels are connected to each other within each segment by two smaller blood vessels. There are also many small capillaries that extend into the skin of the worm, and this is where gas exchange occurs in most species. Annelids do not have a central, well-developed heart, and usually the muscular dorsal blood vessel functions to pump blood through the circulatory system. In some species there may be a number of muscular blood vessels that function as blood-pumping hearts. Earthworms have five of these muscular, heart-like vessels located in the front region of the body. A diagram of a dissected annelid showing part of the closed circulatory system is shown in Figure below.

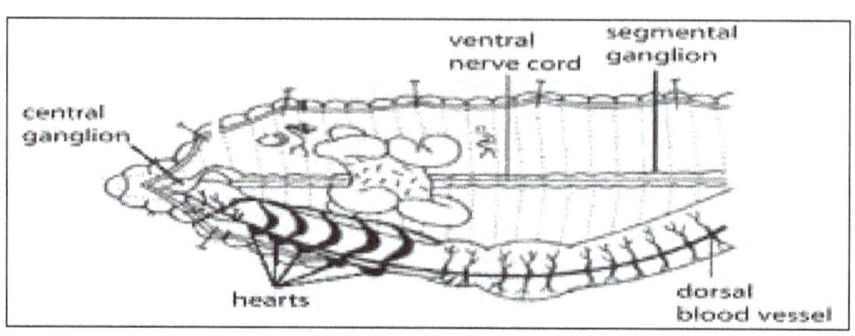

A diagram of the internal anatomy of an annelid: This diagram shows the organization of the closed circulatory system, including the muscular heart-like vessels and the dorsal blood vessel. Notice the smaller blood vessels branching from the dorsal blood vessel. In a closed circulatory system, blood is transported to each tissue inside of tiny blood vessels. It never flows freely within body cavities of the animal. Several components of the annelid nervous system are also depicted in this diagram, including the central ganglion in the head region, the ventral nerve cord, and the many segmental ganglia.

Excretion

The annelid excretory system is made up of long tubular organs called nephridia. Many species have a pair of nephridia in each segment. The nephridia each have an opening called a nephrostome that takes in body fluids from the coelom, and an exit pore in the body wall called the nephridiopore releases waste from the worm. As body fluids travel through the nephridia, both nutrients that are useful to the organism and water are reabsorbed, leaving behind concentrated waste fluid that is excreted through the nephridiopore. The nephridia are lined with short, hair-like projections called cilia that beat repeatedly to maintain the flow of fluid through the organ.

Digestion

The digestive system of annelids shows a lot of variation depending on the feeding habits of the species. All species have a complete digestive tract that begins with a mouth and ends with an anal opening. The digestive system is generally organized in the following order:

- Mouth.
- Pharynx.
- Esophagus.
- Stomach.
- Intestine.
- Anus.

The mouth has a fairly simple structure in species that consume decaying matter or filter feed, whereas predators and blood-sucking parasites have modifications such as jaws with teeth and a tubular sucking organ, mentioned earlier, called a proboscis. On a side note, the proboscis is also the term used for the trunk of elephants, so there is quite a large range of proboscis structures. The pharynx is the compartment that connects the mouth to the esophagus. The digestive tract also has modifications depending on the diet of the species. Earthworms, for example, are deposit feeders. That means that they eat soil and sand, and their digestive systems extract organic matter, such as leaves and plant materials, from the sediment. In order to handle this diet, earthworms have a muscular digestive organ called a gizzard that breaks up the soil so that the organic matter can be extracted. An interesting modification of the digestive system of Hirudinea that helps them to successfully obtain nutrition by blood-sucking is the makeup of their saliva. Leech saliva contains both painkillers and anti-clotting agents. The painkillers prevent prey from noticing that there is a leech feeding on them, and the anti-clotting factors are important to keep the blood flowing out of the wound created from the bite.

Nervous System

Annelids have a central brain located in the head region. The brain is connected to a ventral nerve cord that runs through the segments along the bottom side of the body. Several components of the annelid nervous system are shown in Figure above. Within each segment, there is a ganglion (plural ganglia) that is a concentrated region of nerve tissue located on the ventral nerve cord. Nerve fibers extend into each segment from their ganglion. There is also a pair of ganglia located near the pharynx called the sub-pharyngeal ganglia. These ganglia connect to the brain using special nerve fibers called the circum-pharyngeal connectives. Along with their centralized nervous system, annelids have a number of different sensory organs that help them to determine what is going on in their environments. These organs include the eyes, palps, and antennae located in the head region. Many species also have sensory organs that can detect chemicals present in the environment.

These are called chemosensory organs. The nervous system of active predatory annelid species often has a larger brain and more sensory organs than sedentary, burrowing species. This makes sense because the lifestyle of predatory species requires more detailed sensory input from the environment and greater nervous system function for complex behavioral responses to that input.

Locomotion

Non-sedentary annelids have several means of locomotion. Aquatic polychaetes can use their paddle-shaped parapodia to swim. Longitudinal muscles that run along the body wall from the front to the back are surrounded by a sheath of circular muscles. Annelids crawl or burrow by combining the actions of these muscles and the bristles, or setae, on the outside of the body. The organism first extends forward and then anchors the front end of the body to a solid surface using the bristles. With the front of the body immobilized the worm contracts muscles to bring the rear end closer to the front end. The front end is then released and extended forward again while the rear end becomes anchored so that it does not slip backward as the body extends forward.

Porifera Anatomy

Sponges lack organs and tissue, and all the cells exhibit considerable independence. The sponge is made up of two single-cell-deep layers and an intermediate mesohyl (mobile cells plus extracellular matrix). The outer (sac) layer consists of flattened polygonal cells called pinacocytes. The middle (mesohyl) layer consists of gelatinous protein/carbohydrate material, a range of mobile cells, and a skeleton of calcareous or siliceous spicules, or of elastic proteinaceous fibers called spongin fibers. The inner layer consists of flagelled cells called collar cells, or choanocytes.

The body is permeated by numerous pores called ostia that open into inhalant canals that lead to the feeding chambers, which are made up of choanocytes; here also are large openings, termed oscules, fed by exhalant canals that carry the water current from the choanocyte chambers to the exterior. The concerted whipping action of the choanocyte flagella creates a current of water from ostia through the sponge body oscules. The choanocytes filter plankton and small bits of organic detritus from the water and, like the pinacocytes, absorb oxygen. Food is digested in ameboid archaeocytes that pick up food vacuoles from the choanocytes, which ingest the mainly particulate food. Waste products are carried out through the osculum.

References

- Animal-anatomy, science: factmonster.com, Retrieved 05, Jan 2020
- Amphibian-anatomy: agriculture.gov.au, Retrieved 20, Feb 2020
- Description-and-physical-characteristics-of-reptiles, reptiles: msdvetmanual.com, Retrieved 12, March 2020
- Avian-anatomy-101-390395: thesprucepets.com, Retrieved 15, April 2020
- Fish-are-animals-that-are-scales-breathe-with-gills-they-are-either-single-fins-pelvic-fins, anatomy, fishing -tips, freshwater, fishing: myfwc.com, Retrieved 13, May 2020
- Biology-advanced-concepts: ck12.org, Retrieved 27, May 2020
- Subphylum-vertebrata, molluscs: mesa.edu.au, Retrieved 02, June 2020

Chapter 2

Animal Cells and Tissues

Animal cells are eukaryotic cells with a membrane-bound nucleus, lacking a cell wall. They contain all the necessary structures and constituents required for the survival of the organism. Tissues are a group of connected cells performing similar functions with different tissue groups varying in structure, function and origin. Animal tissues can be classified into various types which are examined in detail in this chapter.

Animal Cell

A cell is the smallest (microscopic) structural-functional unit of life of an organism. The cells that constitute an animal are called Animal cells and those that constitute plants are known as plant cells. Most cells are covered by a protective membrane known as the cell wallwhich gives the cells their shape and rigidity.

An animal cell is a eukaryotic cell that lacks a cell wall, and it is enclosed by the plasma membrane. The cell organelles are enclosed by the plasma membrane including the cell nucleus. Unlike the animal cell lacking the cell wall, plant cells have a cell wall.

Since animal cells lack a rigid cell wall it allows them to develop a great diversity of cell types, tissues, and organs. The nerves and muscles are made up of specialized cells that plant cells cannot evolve to form, hence giving these nerve and muscle cells have the ability to move.

Animal Cell Size and Shape

Animal cells come in all kinds of shapes and sizes, with their size ranging from a few millimeters to micrometers. The largest animal cell is the ostrich egg which has a 5-inch diameter, weighing about 1.2-1.4 kg and the smallest animal cells are the neurons of about 100 microns in diameter.

Animal cells are smaller than the plant cells and they are generally irregular in shape taking various forms of shapes, due to lack of the cell wall. Some cells are round, oval, flattened or rod-shaped, spherical, concave, rectangular. This is due to the lack of a cell wall. Note: most of the cells are microscopic hence they can only be seen under a microscope in order to study their anatomy. But animal cells share other cellular organelles with plant cells as both have evolved from eukaryotic cells.

As noted earlier, animal cells are eukaryotic cells with a membrane-bound nucleus. Therefore they have their genetic material in the form of DNA enclosed in the nucleus. They also have several structural organelles within the plasma membrane which perform various specific functions for proper cell function and generally to maintain the body normal mechanisms.

Types of Animal Cells

There are heaps of different types of animal cells and these are just a few from common tissues like skin, muscle, and blood.

Skin Cells

The skin cells of animals mostly consist of keratinocytes and melanocytes – 'cyte' meaning cell. Keratinocytes make up around 90% of all skin cells and produce a protein called 'keratin'. The keratin in skin cells helps to make skin an effective layer of protection for the body. Keratin also makes hair and nails.

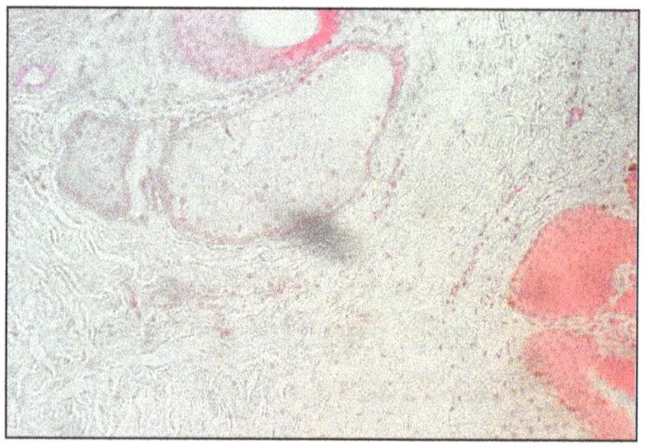

Melanocytes are the second main type of skin cell. They produce a compound called 'melanin' which gives skin its color. Melanocytes sit underneath keratinocytes in a lower layer of skin cells and the melanin they produce is transported up to the surface layers of cells. The more melanocytes you have in your skin, the darker your skin is.

Muscle Cells

Myocytes, muscle fibers or muscle cells are long tubular cells responsible for moving an organism's limbs and organs. Muscle cells can be either skeletal muscle cells, cardiac muscle cells or smooth muscle cell.

Skeletal muscle cells are the most common type of muscle cells and are responsible for making general, conscious movements of the body. Cardiac muscle cells control contractions of heart by generating electrical impulses and smooth muscle cells control subconscious movements of tissues such as blood vessels, the uterus, and the stomach.

Blood Cells

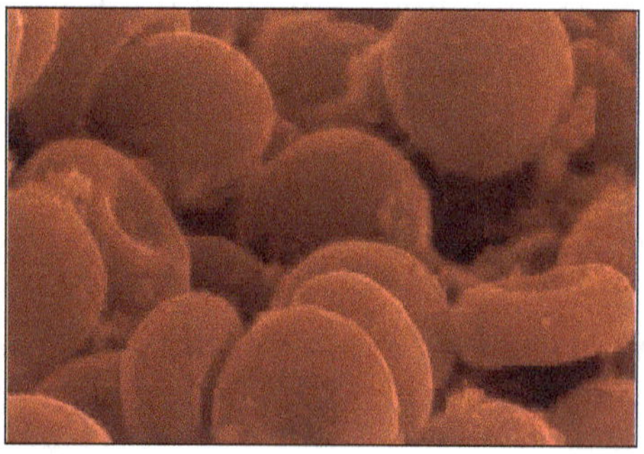

Blood cellsBlood cells can be split into red and white blood cells. Red blood cells make up around 99.9% of all blood cells and are responsible for delivering oxygen from the lungs to the rest of the body. Red blood cells are the only animal cells that do not have a nucleus. White blood cells are a vital part of an animal's immune system and help to battle infections by killing off damaging bacteria and other compounds.

Nerve Cells

Nerve cells, also called neurons, are the main cells of the nervous system. The human brain alone has around 100 billion nerve cells. They are the message carriers of animal cells and deliver and receive signals using dendrites and axons. Dendrites and axons are extensions from the cell that receive and export signals to and from the cell, respectively.

Fat Cells

Fat cells, also known as adipocytes or lipocytes, are used to store fats and other lipids as energy reserves. There are two common types of fat cells in animals – white fat cells and brown fat cells. The main difference between the two cell types is the way they store lipids. White fat cells have one large lipid drop whereas in brown fat cells there are multiple, smaller lipid droplets spread through the cell.

Animal Cell Structure

The animal cell is made up of several structural organelles enclosed in the plasma membrane that enable it to function properly, eliciting mechanisms that benefit the host (animal). The working together of all cells gives an animal its ability to move, to reproduce, to respond to stimuli, to digest and absorb food, etc. Generally, the combined effort by all animal cells is what enables the normal functioning of the body.

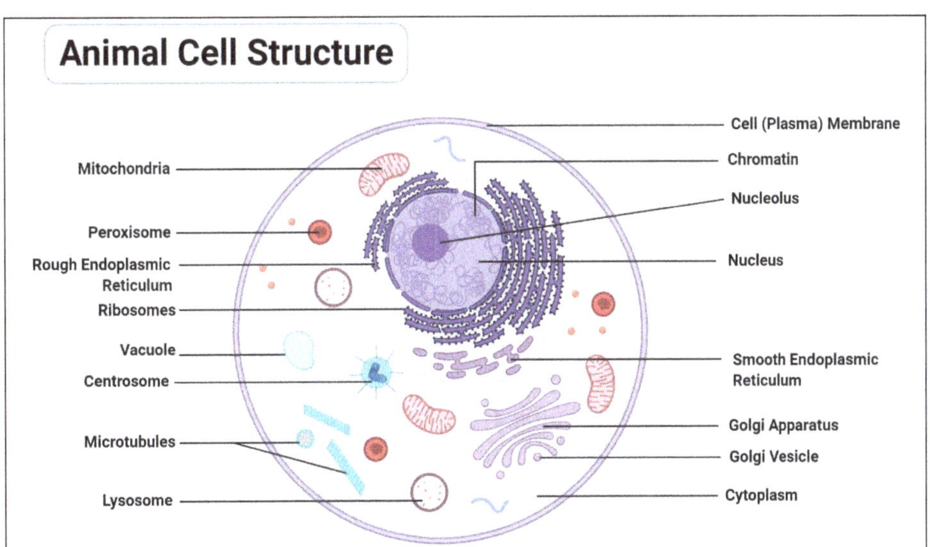

Diagram of Animal Cell.

Animal Cell Organelles

The major cell organelles include:

Plasma Membrane (Cell Membrane)

It is a thin semipermeable protein-membrane layer that surrounds an animal cell.

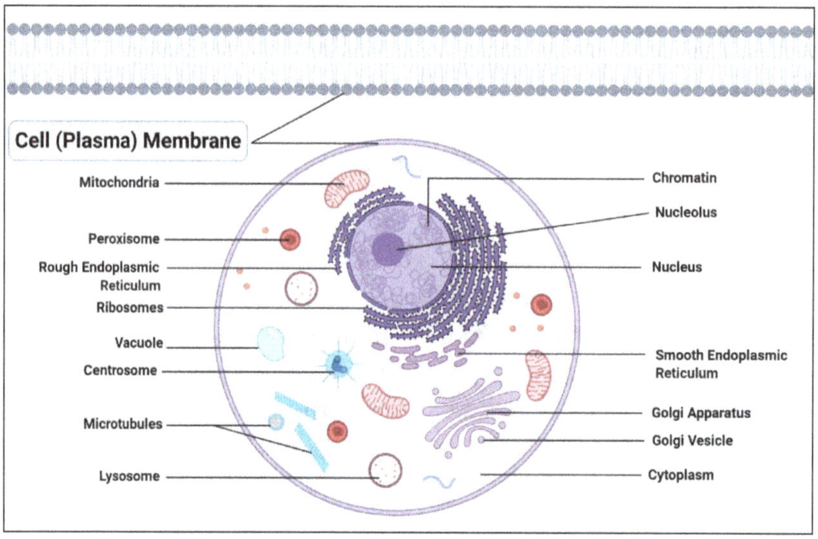

Diagram of Plasma membrane (Cell membrane).

Structure of Plasma Membrane (Cell Membrane)

Thin semi-permeable membrane. It contains a percentage of lipids making a semi-permeable barrier between the cell and its physical environment. It has some protein components a. It is very consistent around the cell. All living cells have a plasma membrane.

Functions of Plasma Membrane (Cell Membrane)

To enclose and protect the cell content. To also regulate the molecules that pass into and out of the cell, through the plasma membrane. Therefore it controls homeostasis. The proteins are actively involved in transporting materials across the membrane. The proteins and lipids allow cell communication, and carbohydrates (sugars and sugar chains), which decorate both the proteins and lipids and help cells recognize each other.

Nucleus

This is a spherical structured organelle found majorly at the center of a cell surrounded by a double-layered nuclear membrane separating it from the cytoplasm. It is held together to the cytoplasm with the help of the filaments and microtubules. It holds other cells organelles including the nucleolus, nucleosomes, and chromatins. A cell has one nucleus which divides producing multinucleated cells e.g. the skeletal muscle cell fibers. Some cells lose their nuclei after maturations e.g. the red blood cells.

The double-layered membrane is a continuous channel of membranous from the endoplasmic reticulum network. The membrane has pores which allow entry of large molecule. Nucleoli (Singular;

nucleolus) are tiny/small bodies found in the nucleus. The nucleus and its component organelles are suspended in the nucleoplasm (House of the chromosomal DNA and genetic materials).

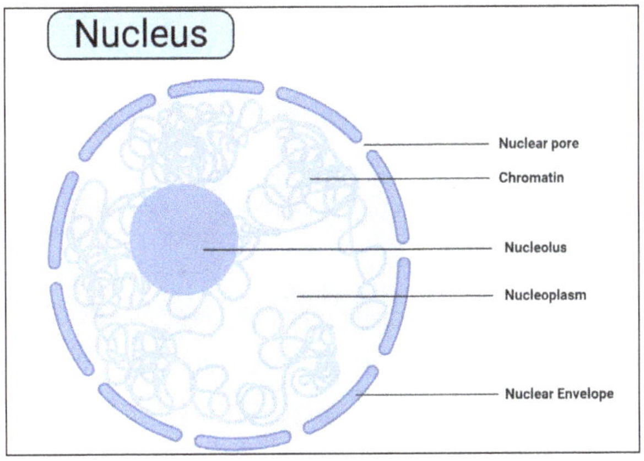

Diagram of Nucleus.

Functions of Nucleus

The primary role of the nucleus is to control and regulate cell activities of growth and maintain cell metabolisms. It also carries the genes that have hereditary information of the cell. The chromosomal DNA and genetic materials, which are made up of genetic coded ultimately make up their proteins' amino acid sequences for use by the cell. Therefore, the nucleus is the information center. It is the site for Transcription (formation of mRNA from DNA) and the mRNA is transported to the nuclear envelope.

Cytoplasm

This is a gel-like material that contains all the cell organelles, enclosed within the cell membrane. These organelles include; Mitochondria, ribosomes, Endoplasmic reticulum, Golgi apparatus, and lysosomes intermediate filaments, microfilaments microtubules, vesicles. This is a gel-like material that contains all the cell organelles, enclosed within the cell membrane. These organelles include; Mitochondria, ribosomes, Endoplasmic reticulum, Golgi apparatus, and lysosomes intermediate filaments, microfilaments microtubules, vesicles.

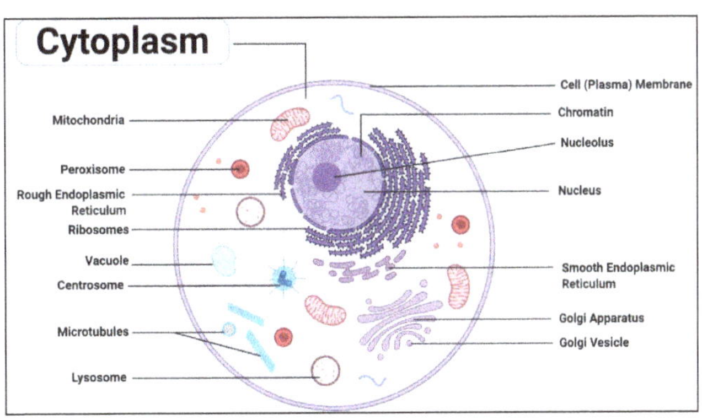

Diagram of Cytoplasm.

Mitochondria

These are membrane-bound organelles located in the cytoplasm of all eukaryotic cells. The number of mitochondria found in each cell varies widely depending on the function of the cell it performs. For example, erythrocytes do not have mitochondria while the liver and muscle cells have thousands of mitochondria.

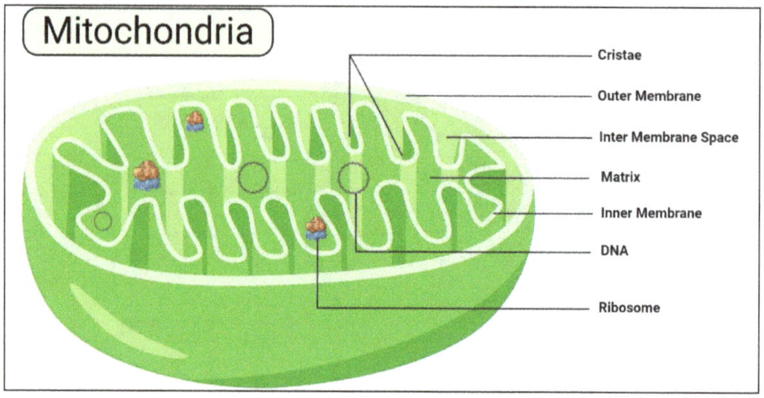

Diagram of Mitochondria.

Structure of Mitochondria

- They are rod-shaped or oval or spherically shaped, with a size of 0.5 to 10 μm. Mitochondria have two special membranes – outer and inner membrane.
- They have a mitochondrial gel-matric in the central mass. The membranes bend into folds known as cristae.

Functions of Mitochondria

Their primary function is to generate energy for the cell i.e. they are the power generators, producing energy in form of Adenosine Tri-phosphate (ATP), by converting nutrients and oxygen into energy enabling the cell to perform its function and to also release excess energy from the cell. Mitochondria also store calcium which assists in cell signaling activity, generating cellular and mechanical heat and mediating cellular growth and death.

The outer membrane is permeable, allowing the transport of small molecules and a special channel to transport large molecules. The inner mitochondrial membrane is less permeable thus allowing very small molecules into the mitochondrial gel-matrix in the central mass. The gel matrix is composed of the mitochondria DNA and enzymes for the Tricarboxylic Acid (TCA) cycle or the Kreb's Cycle.

The TCA cycle uses up the nutrients, converting them into by-products that the mitochondria use for producing energy. These processes take place in the inner membrane because the membrane bends into folds called the cristae, where the protein components used for the main energy production system cells, known as the Electron Transport Chain (ETC). ETC is the main source of ATP production in the body.

The ETC involves several sequences of oxidation-reduction reactions to transport electrons from one protein component to another, thus producing energy that is used for phosphorylation of ADP (Adenosine diphosphate) to ATP. This process is called the chemiosmotic coupling of oxidative phosphorylation. This mechanism gives energy to most cellular activities including muscle movement and the power up the general brain function.

Some if not all proteins and molecules that make up the mitochondria come from the cell nucleus. The mitochondrial nucleus genome has 37 genes of which 13 of these genes produce most of the components of the ETC. However, the mitochondrial DNA is very vulnerable to mutations because they don't possess a large DNA repair mechanism, a common element found in other nuclear DNAs.

Moreover, Reactive Oxygen Species ((ROS)) also called free radicals are produced in the mitochondrion, because of the preference for abnormal production of free electrons. These electrons are neutralized by antioxidant proteins in the mitochondrion. However, some of the free radicals can damage mitochondrial DNA (mtDNA).

Equally, consumption of alcohol can cause damage to the mtDNA because excess ethanol in the body causes saturation of the detoxifying enzymes leading to the production and leakage of highly reactive electrons into the cytoplasmic membrane and into the mitochondrial matrix, combining with other cellular molecules forming numerous radicals that significantly cause cell damage.

Most organisms inherit the mtDNA from their mother. This is because the maternal egg donates most of the cytoplasm to the embryo while the mitochondria inherited from the father's sperm are destroyed. This causes the origin of inherited and acquired mitochondrial diseases due to mutations transmitted into the embryo from the maternal and paternal DNA or maternal mtDNA. Such diseases include Alzheimer's disease and Parkinson's disease. When mutated mtDNA accumulates over time has been linked to aging and the development of certain cancers and diseases. Naturally, mitochondria play a major role in programmed cell death (apoptosis) and due to mutations in the mtDNA can inhibit cell death-causing the development of cancer.

Ribosomes

They are small organelles majorly made up of 60% RNA cytoplasmic- granules and 40% proteins. All living cells contain ribosomes, which may be freely circulating in the cytoplasm and some are bound to the endoplasmic reticulum. It is the site for protein synthesis.

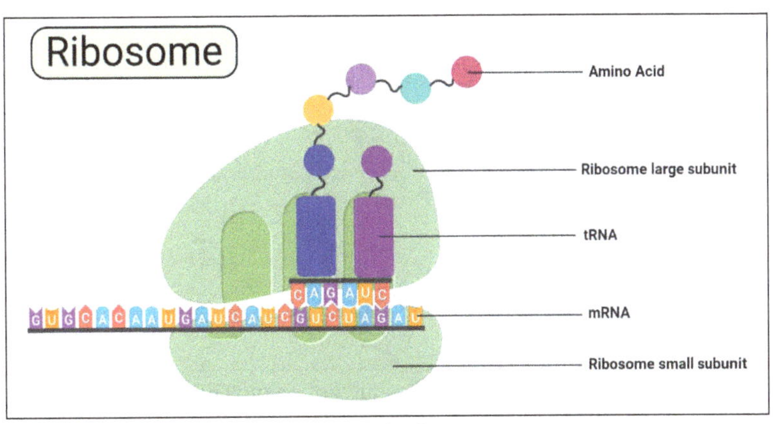

Diagram of Ribosome.

Structure of Ribosomes

Ribosomes are made up of ribosomal proteins and ribosomal RNA (rRNA). In a eukaryotic cell, ribosomes constitute half ribosomal RNA and half ribosomal proteins.

Each ribosome is made up of two subunits i. e large subunit and small subunit with their own distinct shapes. These subunits are designated as the 40s and 60s in the animal cell.

Functions of Ribosomes

Ribosomes that occur as free particles are attached to the endoplasmic reticulum membrane occurring in large numbers accounting for about a quarter of the cell organelles. A single replicated cell has about 10 million ribosomes.

The ribosomal subunits are the site for genetic coding into proteins. On the ribosomes, the mRNA helps determine the coding for Transfer RNA (tRNA) which also determines the protein amino acid sequences. This leads to the formation of the rRNA which is involved in the catalyzation of peptidyl transferase creating the peptide bond found between the amino acid sequences that develop the proteins. The formed proteins then detach from the ribosomes, migrating to other cell parts for utilization by the cell.

Endoplasmic Reticulum (ER)

Structure of Endoplasmic Reticulum (ER)

This is a continuous folded membranous organelle found in the cytoplasm made up of a thin network of flattened interconnected compartments (sacs) that connects from the cytoplasm to the cell nucleus.

Within its membranes, there are membranous spaces called the cristae spaces and the membrane folding is called cristae. There are two types of ER based on their structure and the function they perform including Rough Endoplasmic reticulum and the Smooth endoplasmic reticulum.

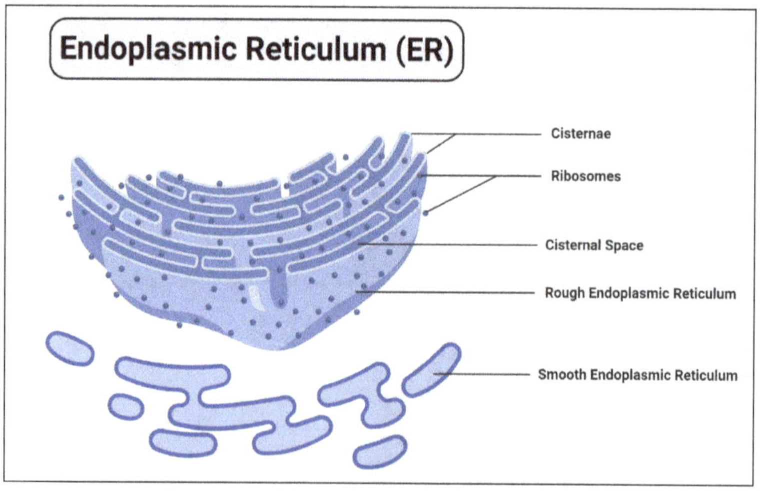

Diagram of Endoplasmic Reticulum (ER).

Functions of Endoplasmic Reticulum (ER)

Manufacturing, processing and transporting proteins for cell utilization both in and out of the cell. This is because it is directly connected to the nuclear membrane providing a passage between the nucleus and the cytoplasm.

The ER has more than half the membranous cell content, hence it has a large surface area where chemical reactions take place. They also contain the enzymes for almost all the cell lipid synthesis hence they are the site for lipid synthesis. The variation in physical and functional characteristics differentiates the ER into two types' i.e. Rough endoplasmic reticulum and Smooth endoplasmic reticulum.

Types of Endoplasmic Reticulum

Rough Endoplasmic Reticulum (Rough ER) – Rough ER is called "rough" because there surface is covered with ribosomes, giving it a rough appearance. The function of the ribosomes on rough ER is to synthesis proteins and they have a signaling sequence, directing them to the endoplasmic reticulum for processing. Rough ER transports the proteins and lipids through the cell into the cristae. They are then sent into the Golgi bodies or inserted into the cell membrane.

Smooth Endoplasmic Reticulum (Smooth ER) – Smooth ER is not associated with ribosomes and their unction is different from that of the rough endoplasmic reticulum, despite lying adjacent to the rough endoplasmic reticulum. Its function is to synthesis lipids (cholesterol and phospholipids) that are utilized for producing new cellular membranes. They are also involved in the synthesis of steroid hormones from cholesterol for certain cell types. It also contributes to the detoxification of the liver after the intake of drugs and toxic chemicals.

There is also a specialized type of smooth ER known as the sarcoplasmic reticulum. Its function is to regulate the concentration of Calcium ions in the muscle cell cytoplasm.

Golgi Apparatus (Golgi Bodies/Golgi Complex)

Structure of Golgi Apparatus (Golgi Bodies)

These are membrane-bound cell organelles found in the cytoplasm of a eukaryotic cell, next to the endoplasmic reticulum and near the nucleus. Golgi bodies are supported together by cytoplasmic microtubules and held by a protein matrix.

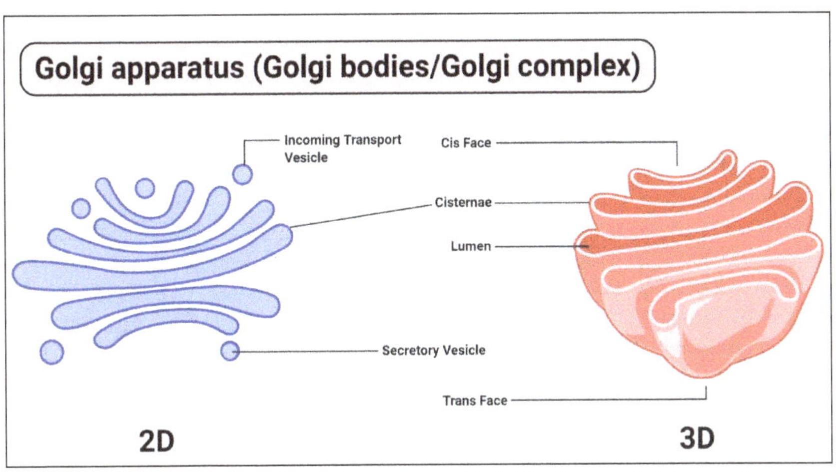

2D and 3D Diagram of Golgi apparatus (Golgi bodies or Golgi complex).

It is made up of flattened stacked pouches known as cisternae. These cisternae maybe 4- 10 in number for animal cell Golgi bodies though some organisms like single-celled organisms have about 60 cisternae.

They have three primary compartments known as cis (Cisternae Nearest the Endoplasmic Reticulum), medial (central layers of cisternae) and the Trans (cisternae farthest from the endoplasmic reticulum). Animal cells have very few (1-2) Golgi bodies while plants have a few hundred.

Functions of Golgi Apparatus (Golgi Bodies)

Their primary function is to transport, modify and pack proteins and lipids into the Golgi vesicles to deliver them to their target sites. Animal cells contain one or more Golgi bodies while plants have a few hundred.

Cis and trans Golgi network make up the outer layer of cisternae at the cis and trans face and they are responsible for sorting proteins and lipids received at the cis face and released by the trans face, by the Golgi bodies.

The cis face collects the proteins and lipids, of fused vesicles in clusters. The fused vesicles move along the microtubules through a specialized compartment known as the vesicular-tubular cluster. This compartment is found between the endoplasmic reticulum and the Golgi apparatus.

The vesicle clusters fuse with the cis Golgi network, delivering the proteins and lipids into the cis face cisternae and as they move from the cis face to the trans face, they get modified to functional units. These functional units get delivered to intracellular and extracellular components of the cell. Modification mechanisms include:

- Cleaving of oligosaccharides chains: Attachment of sugar moieties of different side chains. Adding fatty acids and/or phosphate groups by phosphorylation, and/or removing monosaccharides e.g. the removal of the mannose moieties takes place in the cis and the medial cisternae while adding of galactose takes place in the Trans cisternae.
- Sorting of the modified proteins and lipids occurs in the trans-Golgi network and packed into the trans vesicles, which then delivers them to the lysosomes or sometimes to the cell membrane for exocytosis. Assisted by ligands bound to receptors triggering fusion and protein secretion.

Lysosomes

It is also known as cell vesicles; Lysosomes were discovered by Christian Rene de Duve, a Belgian cytologist in the 1950s.

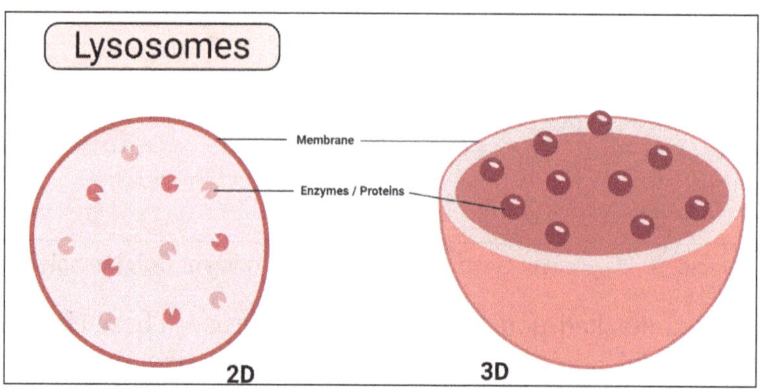

2D and 3D Diagram of Lysosomes.

Structure of Lysosomes

They are round subcellular organelle found in almost all eukaryotic cells. Lysosomes are very acidic organelles containing the digestive enzymes and therefore each of the lysosomes is surrounded by a membrane to protect it from the outer environment.

Functions of Lysosomes

This is the site for digestion of cell nutrients, excretion, and cell renewal. Lysosomes break down macromolecules components from the outside of the cell into simpler elements that are transported into the cytoplasm via a proton pump to build new cell materials.

These macromolecule components include old cells and parts, cell waste products, microorganisms, and cell debris. The digestive enzymes found in the lysosomes are called hydrolytic enzymes or acid hydrolases, breaking down large molecules into smaller molecules that can be utilized by the cell.

These enzymes also break down large molecules e. g proteins, carbohydrates, lipids, into small molecules e.g. amino acids and simple sugars, fatty acids, respectively.

The enzymes are active only on the inside of the acidic lysosome and their acidity protects the cell from degrading itself when there is lysosomal leakage because the cell pH is neutral to slightly alkaline.

Cytoskeleton

Structure of Cytoskeleton

This is a fibrous network that's formed from and by different proteins of long chains of amino acids. These proteins are found in the cell cytoplasm of the eukaryotic cells. They are also made up of 3 types of tiny filaments: Actin filaments (Microfilaments), Microtubules, Intermediate filaments.

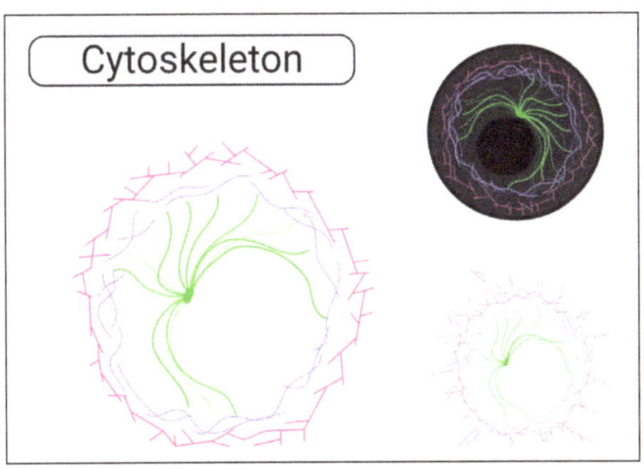

Diagram of Cytoskeleton.

Functions of Cytoskeleton

The cytoskeleton functions to create a network organizing the cell components and to also maintain the cell shape. It also provided a uniform movement of the cell and its organelles, by the filament system network found in the cell's cytoplasm.

It also organizes some of the cell components maintaining the cell shape. It plays a major role in the movement of the cell and some cell organelles in the cytoplasm. The tiny filaments include:

- Actin filaments: Also known as microfilaments; it's a meshwork of fibers running parallel to each other and they play a primary role in giving the cell its shape; they change consistently, helping the cell to move and to also mediate certain cell activities such as adherence ability to substrates and cleavage mechanisms during mitotic cell division.
- Microtubules: These are long filaments that assist in mitosis moving daughter chromosomes to new forming daughter cells.
- Intermediate filaments: They are more stable filaments in comparison to the actin and microtubules. They form the true skeleton of the cell, and the hold the nucleus in its rightful position within the cell. It also allows the cell's elasticity factor enabling it to endure physical tension.

Other proteins that may be added as part of the cytoskeleton of the cell include septin ((assembles the filaments) and spectrin (help maintain the structure of the cell by pulling together the cell membrane with the intracellular surface of the cell).

Microtubules

Structure of Microtubules

These are long, straight, hollow cylinders filaments that are constructed from 13-15 sub-filaments (protofilament) strand of a special globular protein called tubulin, found only in eukaryotic cells. They are found throughout the cytoplasm of the animal cell.

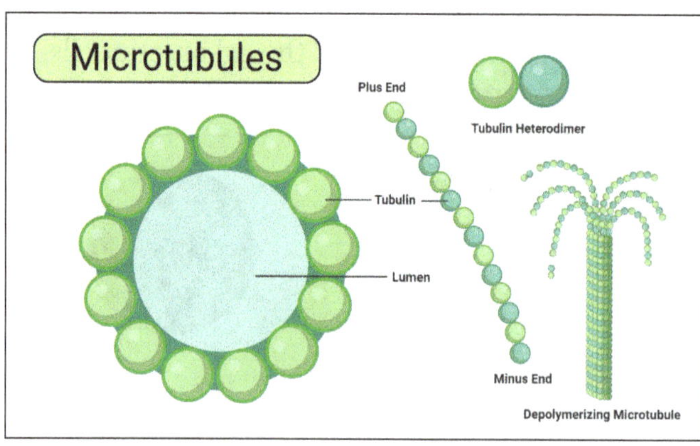

Diagram of Microtubules.

Functions of Microtubules

Transportation of some organelles like the mitochondria and the vesicles i.e. transporting vesicles from the neuron cell body to the axon tips, and back to the cell body.

Structural support, they give characteristic support to the Golgi bodies, holding them within the gel-matrix of the cytoplasm. They provide the rigid and organized component of the cytoskeleton of the cell, enabling a cell to take up a particular shape.

They are the main elements that make up the locomotive projections of a cell (cilia and flagella). They also play a role in forming the spindle fibers of the chromosome of the cell during mitotic cell division.

Centrioles

This is distinctly found in the animal cell, which has the ability to replicate or make copies by itself. It is made up of 9 microtubule bundles and their primary function is to assist in organizing the cell division process.

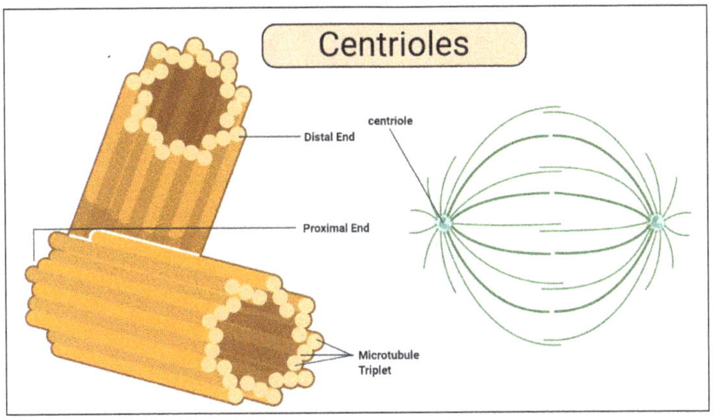

Diagram of Centrioles.

Structure of Centrioles

It is a small structure that is made up of 9 sets of microtubules, placed in groups of three hence they are triplet microtubules. As triplets, they remain very strong together hence they have been observed to be in structures like cilia and flagella.

The triplet microtubules are held together by proteins, giving the centriole its shape. They are found in the centrosome, creating and holding microtubules within the cell.

The triplet microtubules are surrounded by a pericentriolar matrix containing molecules that build up the microtubules. Each microtubule within the triplet microtubule complex is made up of tubulin subunits that join together forming long hollow tubes that look like straw (microtubules).

Functions of Centrioles

The centriole microtubules allow the transportation of substances that are linked together with a glycoprotein to any cell location. The glycoprotein linkage acts as a signaling unit to move specific proteins. The centrioles anchor the microtubules that extend from it and contain the factors needed to create more tubules.

Mitosis is achieved by replication of each centriole which makes duplicates of each centriole (4 centrioles). The newly formed centrioles divide into two centrosomes, each centriole at an angle to the second centriole. The microtubules between the centrosomes push the pairs of centrioles apart, to the opposite ends of the cell. When the centrioles are in place, the microtubules extend to the cell cytoplasm, to seek for the chromosome. The microtubules then bind to the chromosome at the centromere. The microtubules are then unassembled for the centriole moving the chromosomes apart.

Peroxisomes

These are tiny bodies found in the cytoplasm.

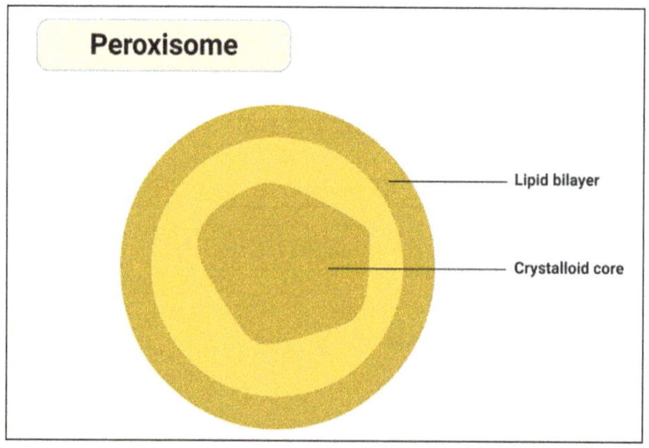

Diagram of Peroxisome.

Structure of Peroxisomes

They are spherically shaped, bound by a membrane and they are the most common micro-bodies in the cell cytoplasm.

Functions of Peroxisomes

- Lipid metabolism.
- Chemical detoxification by moving hydrogen atoms from various oxygen molecules to produce hydrogen peroxide, hence neutralizing body poison such as alcohol.
- Its mechanism in Reactive Oxygen species is highly essential.

Cilia and Flagella

These are locomotive projections found on the surface of the cell.

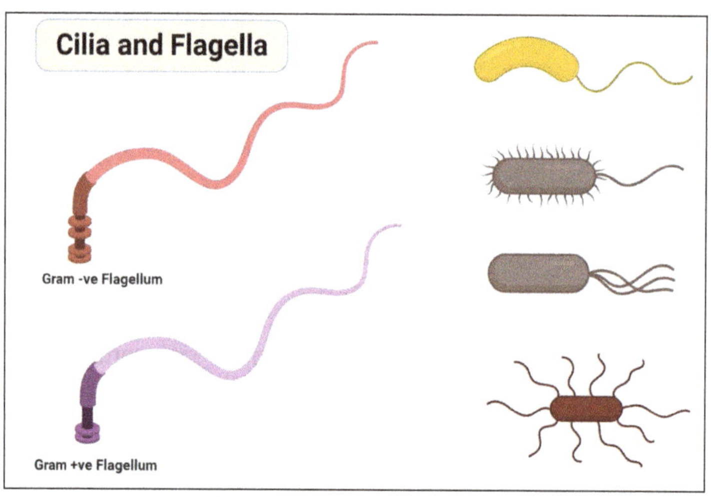

Diagram of Cilia and Flagella.

Structure of Cilia and Flagella

They are made of strands of filaments. these filaments have partial and complete microtubules that extend the projections. Partial microtubules don't extend to the tip of the cilium and the complete microtubules extend to the tip of the cilium.

The microtubules also have motor proteins known as dynein making a link between the partial microtubules to the complete microtubules. The whole collection is combined together as extensions on the plasma membrane of the cell.

Functions of Cilia and Flagella

Sperm cells have flagella allowing it to swim to the ova for fertilization. For single cells, such as sperm, this enables them to swim.

Cilia in the animal cell helps move fluids away from and past immobile cells. Cilia help move surface particles especially on the epithelial lining of the nostrils, and moving mucus over the surface of the cell.

Endosome

These are vesicles bound by membranes and formed by a mechanism of endocytosis. They are found in the cell cytoplasm.

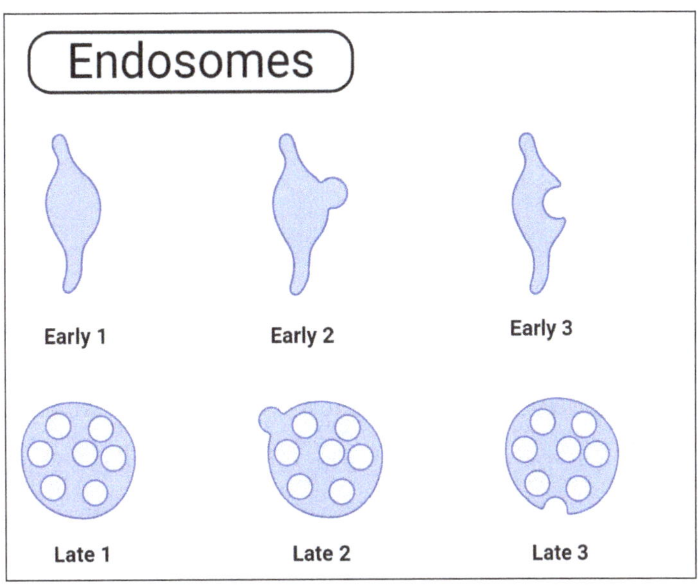

Diagram of Endosomes.

Structure of Endosome

They are membranous organelles that are bound to the cell membrane.

Functions of Endosome

Its main function involves folding in of the plasma membrane. The folding allows diffusing in of molecules through the extracellular fluids. Their primary role is to remove waste materials from the cell by endocytic processes such as exocytosis and phagocytosis.

Vacuoles

These are fluid-filled cell organelles enclosed by a membrane.

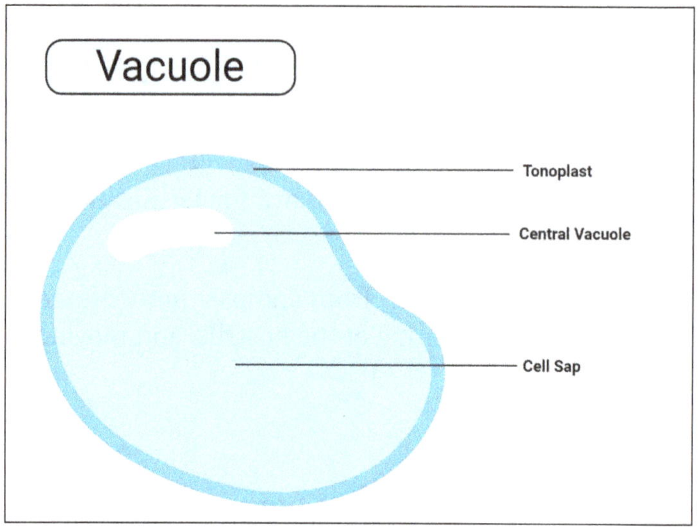

Diagram of Vacuole.

Structure of Vacuoles

They are membrane-bound sacs found within the cell cytoplasm. The vacuole sac has a single membrane surrounding it known as a tonoplast and this membrane resembles the plasma membrane.

Functions of Vacuoles

- Their primary function is to store food, water, carbohydrates in the form of sugars and waste materials.
- Tonoplast is a regulator controlling the inflow and outflow of small across a protein pump.
- Acts as the guard for what kinds of matter are allowed passage to and from vacuoles.
- They also remove toxic substances and waste materials from the cell as a protection strategy.
- They also remove poorly folded proteins from the cell.
- Vacuoles also can be able to change their functionality to provide necessary roles that suit the cell, by being able to change shape and size.

Microvilli

These are surface protrusions found in the intestinal lining, on egg cell surfaces, and on white blood cells.

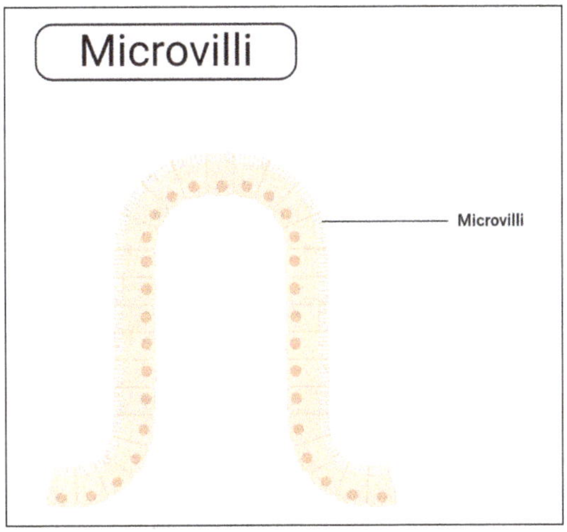

Diagram of Microvilli.

Structure of Microvilli

These are surface protrusions formed from accessory proteins of the actin filaments. The accessory proteins bundle together to form microvilli on the surface of the cell membrane.

Functions of Microvilli

- In the small intestines, they increase the surface area for the absorption of digested food and water. Some microvilli may be found in the ear for detection of sound and they transmit the sound waves to the brain through an electric signal.
- They also help to anchor the sperm to the egg for easy fertilization.
- In white blood cells, they also act as anchors allowing the white blood cells freely moving in the circulatory system to attach to possible pathogens.

Animal Tissue

A tissue is a group of cells that work together to carry out one or several functions in organisms such as plants and animals. These cells communicate and adhere to one another by direct cell-cell contacts and through intermediary molecules like those that form the extracellular matrix. Each organ of the body is formed by several types of tissues. The part of science dealing with plant and animal tissues is known as histology. It is a mostly descriptive discipline that uses light and electron microscopy to describe tissue morphology. The anatomy and organization of tissues are essential to understand the normal physiology and pathology of the organs. Histopathology is a branch of histology that deals with alterations and diseases of tissues.

Types of Animal Tissue

There are four types of tissues found in animals: epithelial tissue, connective tissue, muscle tissue, and nervous tissue.

Epithelial Tissue

Epithelial tissues line body surfaces and cavities, as well as form glands. The cells of the tissue are closely connected to each other via cellular junctions and because epithelium is found on the edges of organs, it has two distinct surfaces. The apical surface is exposed to the body cavity or exterior, while the basal surface is adjacent to the underlying tissue (see the figure below). Epithelia contain no blood vessels (they are non-vascular) and are dependent upon the underlying connective tissue for nutrients.

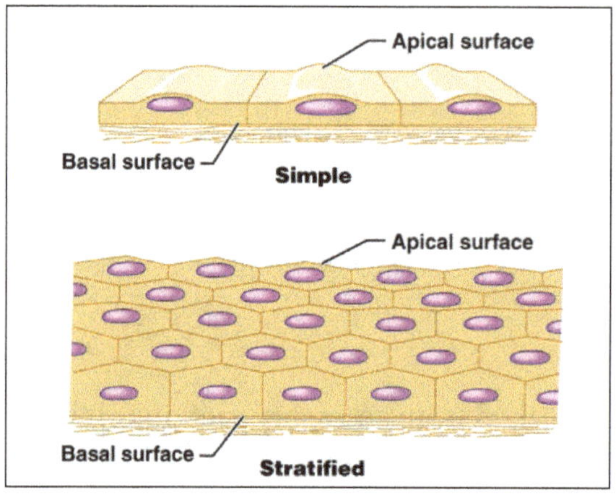

Epithelial cells can be one of three shapes: squamous (flattened), cuboidal, or columnar. If the epithelium consists of one cell layer it is called simple epithelium. Stratified epithelium has more than one layer. The combination of cell shape and the number of cell layers is used to classify epithelial tissue. (I.e. simple columnar epithelium would consist of a single layer of column shaped cells). Cell shape in stratified epithelium can vary, but the shape of the cells at the apical surface determines the name.

The function of epithelial tissue varies depending on its location. As you explore the types of epithelial tissues below, think about how the structure of each tissue fits its function.

Simple Squamous Epithelium

Blood vessels (artery, vein, and nerve): Simple squamous epithelium comprises the inner lining of blood vessels, where it provides a smooth surface that reduces friction as blood travels through the vessels. The blood vessel shows a cross section of an artery and a vein. The wavy lining of the vessel lumen (interior) is simple squamous epithelium.

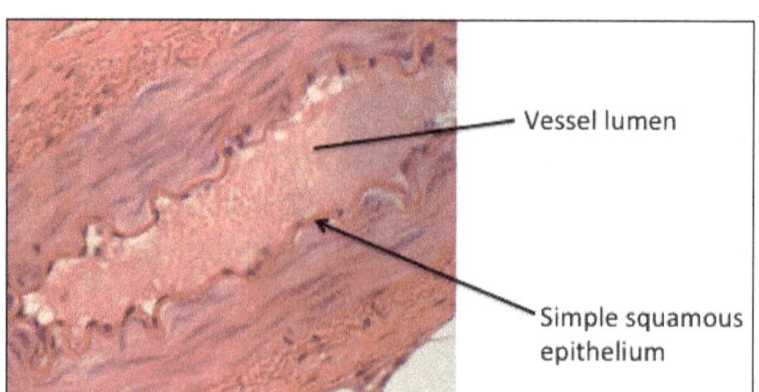

Lung: The walls of lung air sacs (alveoli) are also composed of simple squamous epithelium. Air sacs are the location of gas exchange between the air and blood.

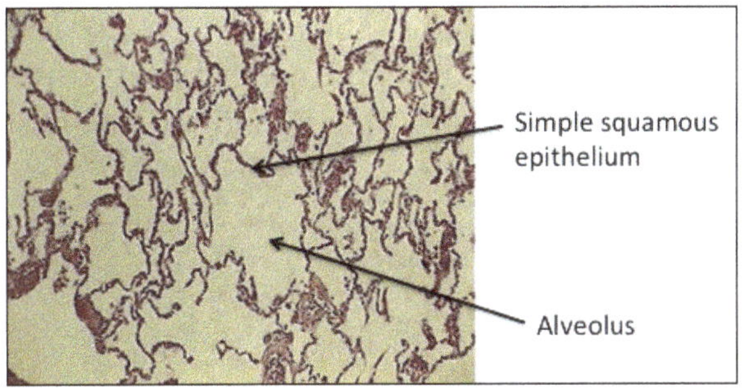

Simple Cuboidal Epithelium

Simple cuboidal epithelium (Kidney): The tubules of the kidney are composed of a single layer of cuboidal cells. The kidney shows cross sections of many tubules, all of which are lined with simple cuboidal epithelium. These cells are active in absorption and secretion of various substances from or into the kidney filtrate (which ultimately becomes urine). Note the shape of the epithelial cells and the centrally located nuclei.

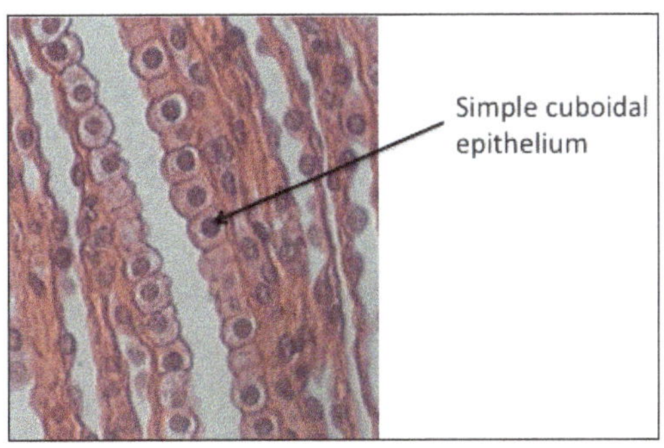

Simple Columnar Epithelium

Duodenum/Small intestine: The intestinal lining is a simple columnar epithelium. The primary function of these cells is absorption of nutrients. The large, oval shaped nuclei that are positioned near the basal edge of the cells. Also note the large, clear goblet cells that are interspersed in the epithelial layer. These are glandular cells that secrete mucus that helps protect the underlying tissues.

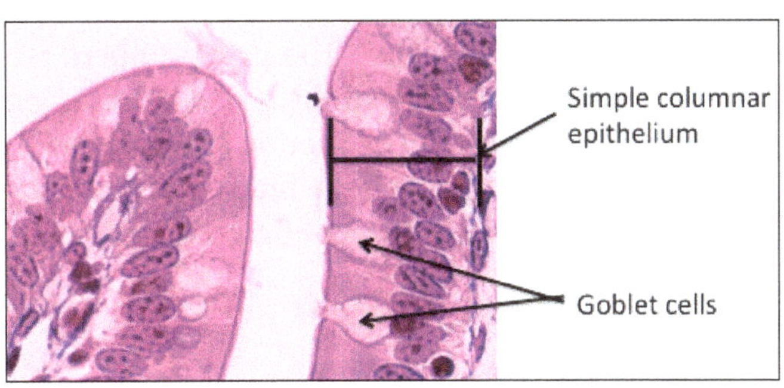

Stratified Squamous Epithelium

Esophagus/stomach: Stratified squamous epithelium consists of multiple layers, with squamous cells at the apical surface. The primary function of this type of epithelium is protection. Areas subject to abrasion, like the mouth, esophagus, and skin, have stratified epithelium. Cells at the apical surface can be scraped away (for instance, by food particles traveling down the esophagus), but the layered nature of the epithelium ensures that the underlying tissues are protected. Note the thick layer of epithelium on the esophagus.

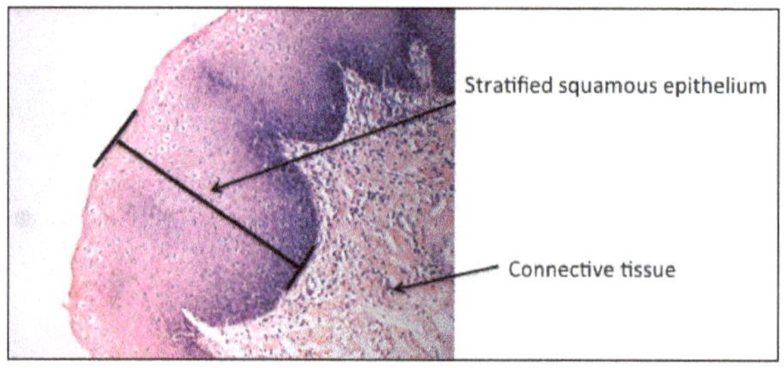

Keratinized Stratified Squamous Epithelium

Palmar Skin (Human skin corpuscle): The epidermis (most superficial layer) of the skin is composed of stratified squamous epithelial cells that contain large quantities of the protein keratin. Keratin is a tough fibrous protein that offers protection from abrasion and water loss. New cells are produced at the basal surface of the epithelium and are gradually pushed towards the apical surface. As they move upwards, they become filled with keratin and eventually die, forming a layer of dead, keratin filled cells on the apical surface of the epidermis. Examine the palmer skin, noting the entire epidermis and the layer of dead cells at the apical surface. The dermis, which lies deep to the epidermis, is composed of connective tissue.

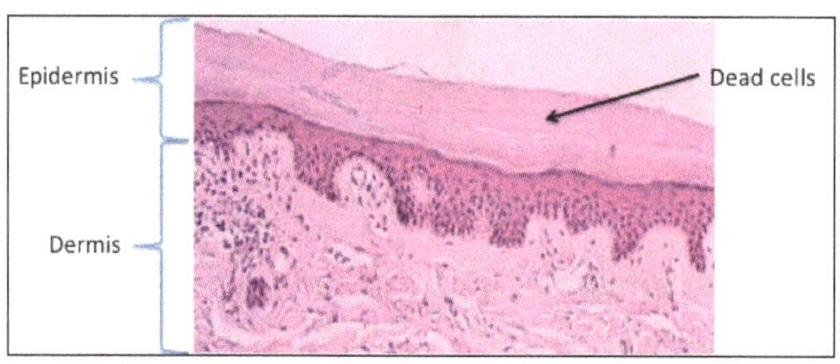

Connective Tissue Connective tissues vary widely in their form and function, but they are all characterized by the presence of extracellular matrix. The extracellular matrix is nonliving material composed of protein fibers and ground substance. The protein fibers are composed of collagen (which gives strength) or elastin (which gives flexibility). The number and type of fibers differs between the various types of connective tissue. The ground substance fills the spaces between the cells and the fibers. It contains interstitial fluid (tissue fluid) and large polysaccharide molecules. The consistency of the ground substance can vary from liquid to gel-like to a solid.

Dense Connective Tissue Palmar Skin

Human skin corpuscle: The layer of skin that lies deep to the epidermis is called the dermis and is composed of dense connective tissue. This tissue contains densely packed bundles of irregularly arranged collagen fibers. It is found in areas of the body that are subject to tension from many different directions. Note the thick layer of dense connective tissue that lies deep to the epithelium on the skin. Nuclei of the connective tissue cells are scattered throughout the collagen fibers.

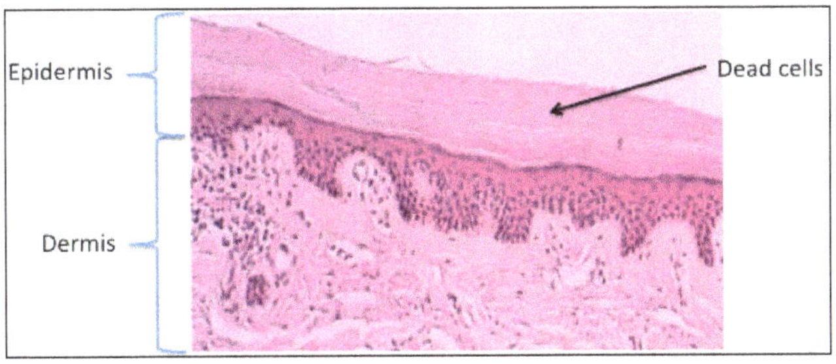

Adipose tissue: Adipose tissue consists of adipocytes, or fat storage cells. It functions in energy storage, insulation, and cushioning. Small pockets of adipose tissue can be found all over the body, but accumulates under the skin (subcutaneous fat) and around certain organs, such as the kidneys. Unlike other connective tissues, it has very little matrix and the cells are closely packed together. Each cell contains a large fat droplet, which pushes the nucleus to the side.

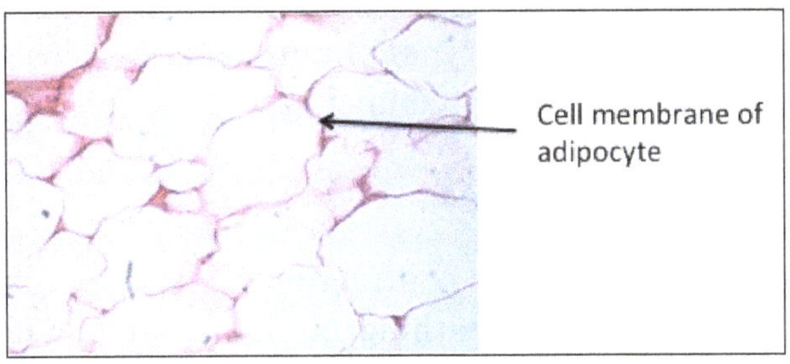

Hyaline cartilage: Hyaline cartilage is the most abundant type of cartilage in the body and is found in the rib cage, the nose, the trachea, and the ends of long bones. It provides structural support (but is more flexible than bone) and has cushioning properties. Hyaline cartilage has a firm matrix with abundant collagen fibers, but the individual fibers cannot be seen under the microscope. When viewed under the microscope the matrix an amorphous quality (no discernable structures). The cells, which are known as chondrocytes, reside in small cavities within the matrix called lacunae.

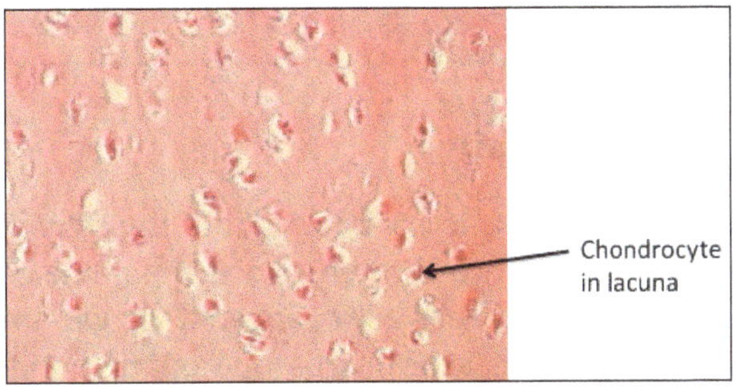

Bone tissue: Bone tissue forms the skeletal system. It functions in structural support, protection, and mineral (calcium) storage. The extracellular matrix of bone tissue contains abundant collagen fibers as well as a hard, calcified ground substance. Mature bone cells, called osteocytes, reside in cavities within the matrix called lacunae. As bone tissue is formed, channels remain in the hardened matrix that provides passageways for blood vessels and nerves. The larger channels are called central canals (Haversian canals). Bone tissue forms in rings (lamellae) around these canals, creating a structure called an osteon. Examine the bone tissue, noting the osteons with their lamellae and bulls-eye like central canals. The lacunae, which contain the bone cells, are visible as small dark patches in the lamellae.

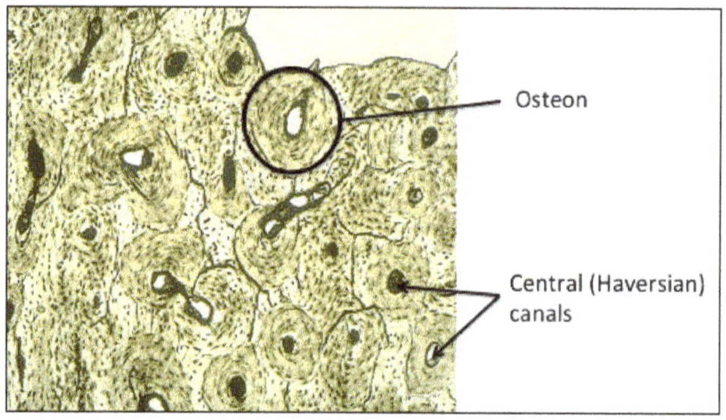

Muscle Tissue

Muscle tissue is specialized for contraction. The cells are elongated, and are also known as muscle fibers. They contain the contractile proteins actin and myosin, which interact to shorten and elongate the cells. There are three different types of muscle tissue: skeletal, cardiac, and smooth. Examine each type of tissue using the muscle composite.

Skeletal Muscle (Muscle Composite)

Skeletal muscles are attached to bones, and contraction of these muscles generates body movements (limb movement, jaw movement, breathing, etc.). The skeletal muscle fibers are long and cylindrical, with multiple peripherally located nuclei. The cells have striations, alternating light and dark bands that result from the ordered arrangement of actin and myosin within the cell.

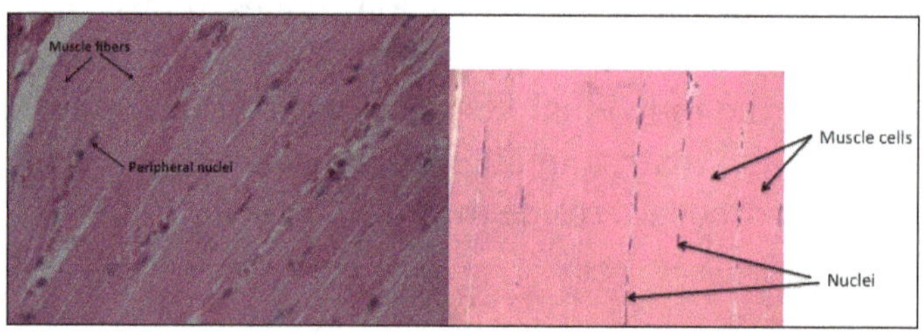

Cardiac Muscle (Muscle Composite)

Cardiac muscle is present in the heart. Cells are striated, but the striations are much less obvious than in skeletal muscle tissue. The cells are shorter than skeletal muscle fibers, have a single nucleus and are often branched. Individual cells are connected via gap junctions and desmosomes. These cellular connections are visible under the microscope as dark bands called intercalated disks. These cellular communication junctions are necessary for the coordinated beating of the heart.

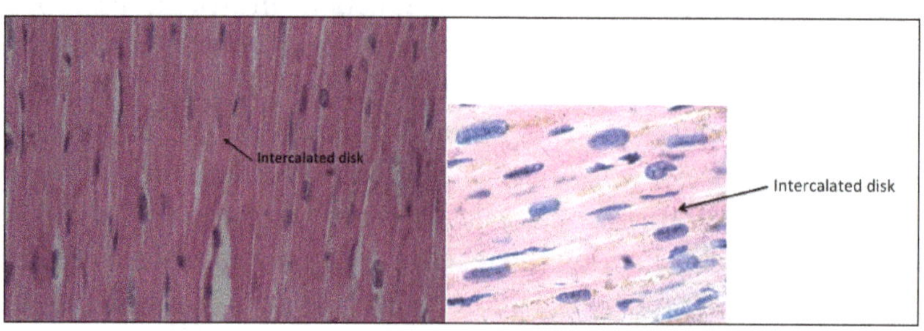

Smooth Muscle (Muscle Composite and Artery/Vein/Nerve)

Smooth muscle tissue is found in the walls of hollow organs, such as the gastrointestinal tract, blood vessels, and the urinary bladder. Contractions of these muscles propel fluid or materials through the organs (i.e. food through the GI tract, blood through blood vessels, urine pushed out of bladder). Smooth muscle cells are not striated (hence the name "smooth" muscle); they have a single nucleus, and have tapered ends. Examine the smooth muscle on the muscle composite as well as the blood vessel. In blood vessels there is a layer of smooth muscle deep to the epithelial layer. It is thicker on the artery than on the vein, but can be seen in both.

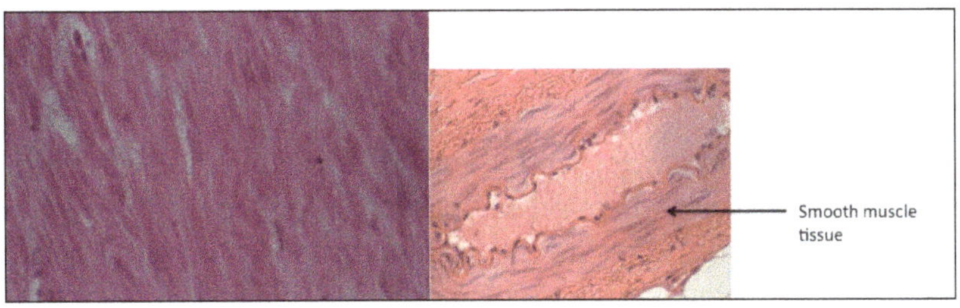

Nervous Tissue

Nervous tissue is specialized for communication and composes the brain, spinal cord, and peripheral nerves. The tissue consists of two major cell types: neurons and glial cells. Neurons communicate with each other via electrical and chemical signals. They have nucleated cell bodies and two types of elongated cellular processes: dendrites – which receive signals, and axons – which send signals.

Glial cells are the support cells of nervous tissue. There are several different types with various functions, including maintaining proper ion concentrations in the fluid surrounding neurons, generating myelin (an insulating material that surrounds some axons), and cleaning up debris. Note the large neurons with their elongated cellular processes and the smaller, more numerous glial cells.

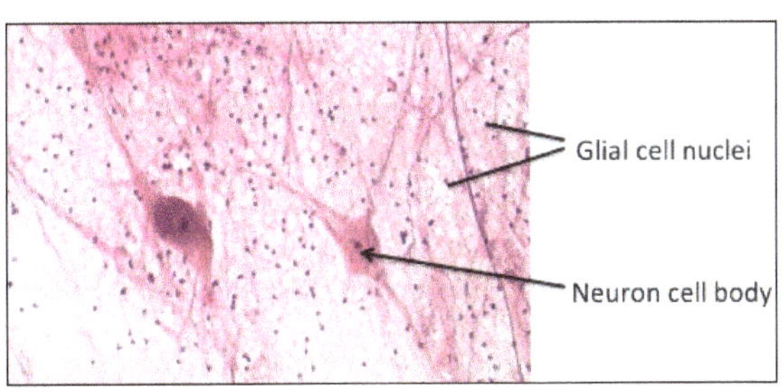

References

- Animal-cell-definition-structure-parts-functions-and-diagram: microbenotes.com, Retrieved 17, June 2020
- The-skin-cells-of-animals-also-makes-hair-and-nails, animal-cells, cells: basicbiology.net, Retrieved 28, July 2020
- Animal-cell-size-and-shape, animal-cell-definition-structure-parts-functions-and-diagram: microbenotes.com, Retrieved 07, August 2020
- Guiada-a-inicio: mmegias.webs.uvigo.es, Retrieved 25, August 2020
- Animal-tissues, biology: augusta.edu, Retrieved 19, Jan 2020

Chapter 3

Systems of the Animal Body

The animal body comprises of many organ systems that help in the everyday functioning of the organism. The muscular system, integumentary system, skeletal system, digestive system, respiratory system, excretory system, reproductive system, endocrine system, etc. are a few of the systems present in all animal bodies. In this chapter, each topic is discussed in detail for the benefit of the reader.

A system is a group of organs that work together and provide an organism with an advantage for survival. It is the most complex organization in your body and the final level of the progression from cells to tissues to organs and then systems. Systems work alone and with other systems to allow your body to maintain homeostasis. Homeostasis is a stable internal environment that allows you (and your cells) to survive.

The major organs in higher animals include the brain, heart, lungs, kidneys, liver, stomach, intestines, and skin, as well as the reproductive organs.

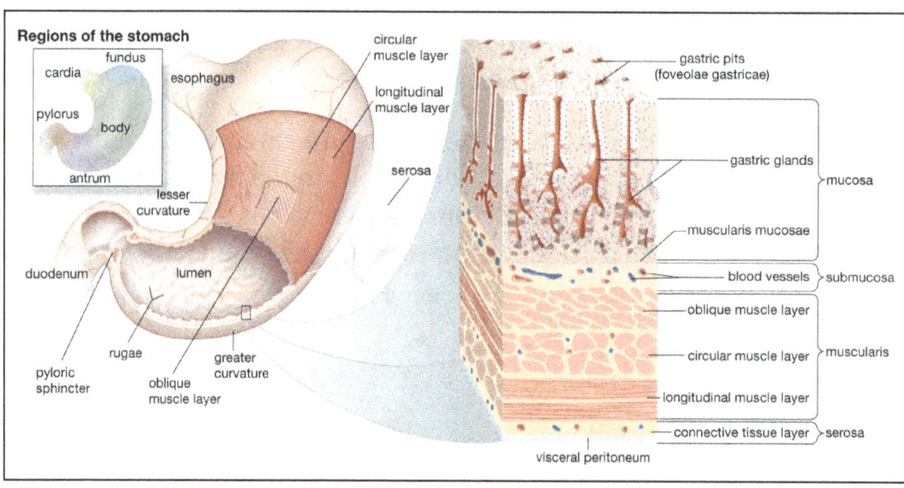

The brain is the control center of the body, directly and indirectly regulating all vital functions, including breathing and the beating of the heart. The heart serves as a pump that circulates blood throughout the body. The lungs function in gas exchange; they are responsible for adding oxygen to and removing carbon dioxide from the blood. The kidneys maintain water balance and help rid the body of metabolic wastes. The stomach and the intestines are the primary sites for the digestion of food and the absorption of nutrients.

One of the most important organs is the liver. Among its many functions are removing wastes and toxic materials from the blood, synthesizing proteins needed for blood clotting, and storing glycogen and vitamins.

Like the liver, the skin has a variety of important functions. It is the main barrier between the inner body and the environment and is the body's first line of defense against "foreign" invaders, such as allergens and infectious organisms. Skin also functions to some extent in gas exchange, especially in amphibians.

All animals have reproductive organs, though these vary in complexity depending upon the animal. In humans and other mammals, the female reproductive organs include the uterus and ovaries; the primary reproductive organs in males are the testes and the penis.

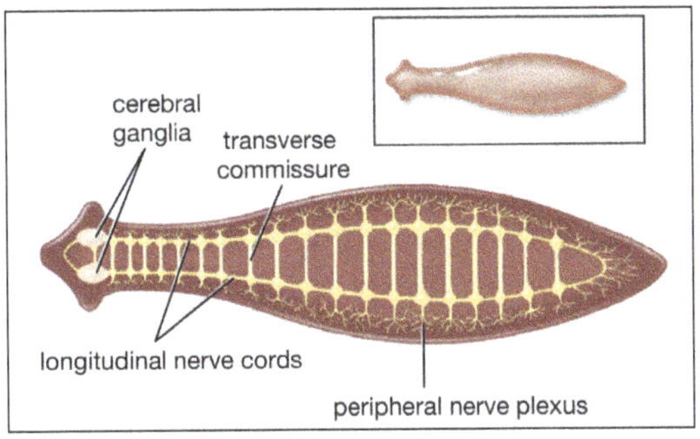

Lower animals have many of the same organs as those found in higher animals, though the organs may exhibit differences in structure or complexity. For example, the brain in many insects and other invertebrates may consist of several simple nerve clusters, or ganglia.

Organs Working Together

Organs are a part of every system. Your heart is classified as an organ and it is a part of the circulatory system. Organs can work within several systems of your body. Many organs also have specific cells or tissues that have different functionality. Your kidneys are not only a part of your excretory system; they also have specific parts that serve the endocrine system.

You, and many advanced mammals, have similar organs and systems. However, there is a wide variety of organ types found throughout the animal kingdom. Some aquatic animals have organs that remove salts from salt water and an animal like a cow might have multiple stomachs in the digestive system.

Systems Can't Work Alone

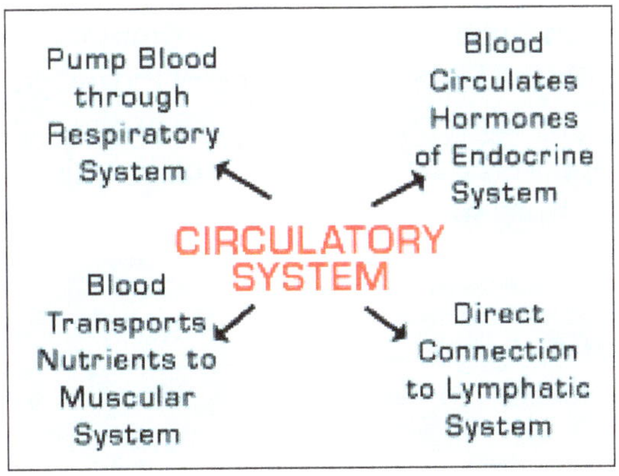

Systems connect to other systems We just explained how organs could be a part of several systems. Similarly, systems rarely work alone. All of the systems in an organism are interconnected. A simple example is the connection between the circulatory and respiratory systems. As blood circulates through your body, it eventually needs fresh oxygen (O2) from the air. When the blood reaches the lungs, part of the respiratory system, the blood is re-oxygenated. Your stomach, part of the digestive system, constantly interacts with your endocrine system and spreads hormones throughout your body.

Examples of Systems

It's easy to point out a few in your body. The two you think of the most are probably your respiratory and digestive system. A couple of times a day you might get hungry, sit down, and have a nice meal. All of that food gets broken down in your digestive system so that your body has energy to survive. The stuff your body doesn't need is pooped out at the other end.

Since you're breathing all of the time, the respiratory system is always at work. You breathe in and out from your nose and mouth while your lungs are the main organs that allow your body to absorb the oxygen you need from the air. There are many other systems in your body and specialized systems in other animals around the world.

Integumentary System

The integumentary system is an organ system that forms the protective covering of an animal and comprises the skin (including glands and their products), hair coat or feathers, scales, nails, hooves and horns.

The integumentary system has a variety of functions; in animals, it serves to waterproof, cushion and protect the deeper tissues, excretes waste, regulates temperature and is the location of sensory receptors for pain, pressure and temperature. Generally mammalian skin is covered with hair and is termed hirsute skin. Where hair is absent, it is termed glabrous skin.

The integumentary system is often the largest organ system. It distinguishes, separates, protects and informs the animal with regard to its surroundings. Small-bodied invertebrates of aquatic or continually moist habitats also respire using the outer layer (integument). This gas exchange system, where gases simply diffuse into and out of the interstitial fluid, is called integumentary exchange. The skin is an organ that shows complex adaptations many of which are species specific - fish, birds, reptiles and amphibians have modifications of the integumen that are tailored to meet the survival needs of each species. Specific lifestyles such as those including hibernation require additional modifications to support the skin such as the phsiologically adapted brown adipose tissue which generates body heat from specialised subcutaneous fat cells. The skin can be reflective of the health status of an animal, being sensative to certain hormonal and nutritional conditions.

Invertebrate Integuments

Components

The integuments of unicellular organisms comprise the cell membrane and any coating it secretes. Almost all bacteria have an external cell wall that maintains the shape and turgidity of the cell and affords protection. Multicellular invertebrates, however, have a single outer layer of epithelial cells, and these may secrete a variety of surface coatings, ranging from the mucous coat of cnidarians (polyps, sea anemones, and jellyfish) to the tough cuticle of insects. The epithelial layer may include cells of several types, such as sensory cells, gland cells, and stinging cells, and the outer surfaces may bear microvilli, cilia, or other fine projections. In addition, the cells may produce excrescences such as bristles, spines, knobs, and ridges.

The firmness of some invertebrate animals, such as annelids (earthworms, marine worms, and leeches) and certain mollusks, depends on the distension by water of the individual cells that form the body wall. In many other forms rigid skeletal materials are deposited either within the cells or on the outer surface. The noncellular coatings of invertebrate integuments are exceedingly varied in composition and extent, and they cut across taxonomic categories. These rigid structures may provide no more than protective armour, but in the arthropods, including crustaceans, insects, and spiders, a multilayered and hardened integument forms an exoskeleton to which muscles are attached. In the echinoderms, the exoskeleton lies below the epidermis.

Differences among Invertebrate Groups

Protozoans

The secreted coatings of protozoans exhibit all grades between soft forms (as in Amoeba) and forms with a cuticle that may be proteinaceous (as in Monocystis) or composed of cellulose (as in the plantlike flagellates). Other protozoans have definite shells, composed of protein incorporating various foreign bodies, such as siliceous plates or calcium carbonate (in most foraminiferans), or cellulose (in the resting stages of slime molds). The radiolarians have an internal lattice of silica that is laid down inside the cell—a kind of internal skeleton, or endoskeleton.

Sponges

Sponges have a simple epithelium, known as the pinacoderm, which both covers the external surfaces and lines the internal waterways. Some sponges deposit needlelike spicules of calcium carbonate in the jelly (mesoglea) beneath this outer epithelium.

Cnidarians

In the cnidarians the epidermis provides all the basic features of an integument. It may contain not only epithelial cells, some of which may be contractile, but also gland cells, pigment cells, stinging cells, and sensory cells with projecting hairs. The outer surface often bears flagella or microvilli. The surface secretion may help in capturing food, adhering to substrates, cleaning away settling debris, or providing support and protection. Some hydrozoans produce a horny covering for the polyps, and others have an external skeleton that is calcareous (containing calcium carbonate). Anthozoans show the same diversity. In the common reef-building corals, the calcareous skeleton is secreted by a part of the epidermis that forms a basal disk. This secretory process is continuous, and the polyp raises itself progressively upon a constantly growing stem of calcium carbonate.

Flukes and Roundworms

The parasitic flukes have a relatively thick integument, which bears many spines and sensory papillae, an apical membrane that is thrown into ridges and pits, and microvilli. The outer part of the integument contains secretory bodies, which are continuously released at the surface to renew the apical membrane. This appears to be a protective device for the parasite related to the immune reaction of the host. Roundworms have a thick, flexible cuticle, with three distinguishable zones, covered by an epicuticle.

Annelids

Annelids have a thin, horny cuticle pierced by pores through which epidermal glands secrete mucus. In some marine annelids, glands are also present that secrete materials constituting a parchmentlike or calcareous tube within which the worm dwells. Earthworms and leeches secrete cocoons from a specialized epidermis in a region of the body known as the clitellum. A major feature of all annelids except leeches is the possession of bristles, or chaetae, of which there are many varieties. The bulk of each chaeta is secreted by a single cell, though the surrounding lateral cells may contribute materials that bring about its hardening.

Arthropods

The exoskeleton attains its most elaborate forms in the arthropods (for example, crustaceans and insects). The insect epidermis lies on a basement membrane and secretes a tough cuticle, the bulk of which is composed of fibres of a material known as chitin embedded in a matrix of protein. Peripheral to this is an epicuticle. Chitin is a high-molecular-weight polysaccharide containing amino groups. It is synthesized within the epidermis from sugars and amino sugars.

In the integument of caterpillars chitin forms a cuticle that is tough but flexible. But in most arthropods the segments of the body or of the limbs are in the form of rigid plates that form a true exoskeleton linked to adjacent segments by flexible membranes. Such cuticles are hard and may be dark in colour. They are said to be tanned, or sclerotized, and in some species they are also mineralized.

Sclerotization involves the molecular stabilization of the protein chains of the cuticles by establishment of cross-links. Sclerotin, the product of sclerotinization, is a kind of natural plastic. In its horny consistency it closely resembles keratin; both are cross-linked, or polymerized, proteins, but the chemical nature of the linkage is different in the two substances. It is probable that other skeletal proteins in invertebrates, such as the spongin of sponges and the conchiolin of mollusks, are also tanned proteins allied to sclerotin.

In many crustaceans—crabs and lobsters, for example—much of the cuticle is rendered hard by the incorporation of calcareous substances such as aragonite or calcite. But sclerotin is actually harder than calcite, and those parts of crustaceans that need to be of maximum hardness, such as the mandibles and the tips of the claws, are in fact composed of sclerotin.

Besides functioning as a skeleton, the cuticle of terrestrial arthropods must act as a waterproof covering in order to prevent these small animals from drying up. This waterproofing is effected by the secretion of a layer of wax on the surface of the cuticle. Such a wax layer, if exposed in an unprotected state, would be excessively fragile. It is commonly protected by a thin layer of a cementlike substance that is poured over its surface by small dermal glands.

The cuticle of arthropods, pierced by ducts of dermal glands that pour out secretions over the surface, is a living structure; it can produce tactile bristles, pigment-bearing scales, claws, wings, and other structures. In some insects it shows brilliant metallic colours that result from the presence of multiple thin plates or ridges in the cuticle. In order that the arthropod may grow, the old cuticle is shed from time to time after a new and larger cuticle has been laid down beneath it. This process is termed molting, or ecdysis. During the time when the new cuticle is hardening, the arthropod is in a very vulnerable condition.

Molting in insects is hormonally controlled. A molting hormone, known as ecdysone, is mainly a product of the thoracic glands, and its secretion is influenced by a prothoracicotropic (or ecdysiotropic) hormone produced by certain cells of the brain. Larvae also possess a juvenile hormone, which decreases in concentration until the imago (adult) emerges. Crustaceans and spiders possess analogous hormones, though their systems are not identical.

Mollusks

The epidermis of mollusks is capable of a variety of functions. Ciliated epithelium is of particular importance for feeding in bivalves and for the gliding movement of snails. Abundant gland cells secrete mucus, which protects the animal from predators and from desiccation. Complex glands secrete the quinone-tanned proteins of the byssus threads, by which mussels anchor themselves, and of the operculum, with which some sea snails stopper their shells. The secretion of pallial glands enables the date mussel to bore into calcareous rock. Some cephalopods (squids, cuttlefish, octopuses) have luminous glands, although it is disputed whether the luminous material is produced by the epithelium itself or by bacteria. Cephalopods also have pigment cells that can be expanded by muscle contraction and can change colour very rapidly.

The shell of mollusks is secreted by the epithelium of the mantle and consists of an outer layer of the horny substance conchiolin, an intermediate prismatic layer composed of calcite, and a smooth inner layer (the nacreous layer) also composed mainly of calcium carbonate. The first two layers are secreted by a marginal band of cells, so that the shell grows at its outer edge. The nacreous layer is secreted by the general surface of the mantle and is the material of which pearls are formed around foreign bodies introduced into the mantle cavity.

Echinoderms

The echinoderms are characterized by a calcareous exoskeleton, which may be a rigid armour, as in echinoids (sea urchins), or of a leathery consistency, as in holothurians (sea cucumbers). The epidermis lies outside of this skeleton. The apical plasma membrane is capable of taking up dissolved organic molecules from the surrounding seawater in amounts that are at least enough for the nourishment of the epidermal layer. Many sea urchins have projecting spines on which the epidermis is worn away to expose the calcareous material.

The Vertebrate Integumentary System

Skin Layers

In all vertebrates the skin has two major layers. The outer, relatively thin epidermis is composed of closely packed cells with little intercellular material; it provides the barrier against attack by chemicals, radiation, or microbes. The underlying dermis (cutis, corium) is thicker and tougher, and its bulk is formed by extracellular materials manufactured by scattered cells. One of its major functions is physical protection. The sensory functions of skin are shared by both epidermis and dermis. The various structures or appendages such as scales, feathers, claws, glands, and hair follicles are derived from either layer or from both. There are considerable differences between the skins of different vertebrate classes, and they are closely related to the environments of the various groups.

The epidermis is the product of the deepest layer of its cells, those that lie immediately over the dermis. From this generative layer, known as the stratum germinativum, cells move outward and become progressively flattened. The surface cells of terrestrial vertebrates, mere remnants of once living cells, are scaly and compressed; they constitute the horny layer, or stratum corneum. The cell fragments of the stratum corneum are composed largely of keratin, a tough insoluble protein. In most land vertebrates the stratum corneum is shed or molted, either periodically and in large fragments or sheets, as in reptiles, or continuously in small patches or scales, as in mammals.

The dermis, which is best developed in mammals, consists largely of fibrous connective tissue (composed of collagen fibres), blood and lymph vessels, smooth muscle cells, and nerve endings. It gives rise to so-called membrane bones—the bony scales of fishes, the bony plates in certain reptiles and mammals, and the membrane bones of the vertebrate skull. Through its blood network the dermis supplies nourishment to the overlying epidermis.

Skin Derivatives and Appendages

Among the notable changes that have taken place during the course of evolution is the development in vertebrates of a variety of glands, pigmentary structures, scales, claws, nails, horns, feathers, and hairs as adaptations to their changing environments.

Skin Glands

The glands of the skin are all exocrine, that is, they secrete their products, usually through ducts, to the epidermal surface. They may be unicellular, as are the goblet cells of fishes, or multicellular, as are the sweat glands of humans. Some multicellular glands are tubular and extrude their secretion into a central space or lumen; some, like the oil-producing sebaceous glands of mammals, form their product by complete breakdown of the cells, a method of secretion known as holocrine. Glands may consist of tubes or sacs, and they may be singular, clustered, or branched; some even contain units of more than one type. They may secrete their product continuously, periodically, or only once.

Mucous glands secrete a protein called mucin, which with water forms the substance known as mucus; this slimy material serves to lubricate the body, thus lessening friction and aiding locomotion in swimming animals. Serous glands produce a watery secretion; sweat glands of mammals are of this type. Sebaceous glands secrete oil, ceruminous glands secrete wax, mammary glands secrete milk, poison glands secrete various toxins, and scent glands secrete a variety of odoriferous substances. Further, certain epidermal glands may be modified into light-producing structures called photophores, seen in the skin of many deep-sea fishes.

Pigment Cells

In fishes, pigment is produced in branched cells known as chromatophores, which can be found in both epidermis and dermis. Rapid colour change, by which some fishes can adapt to a change of background, is brought about by redistribution of the pigment within the cell boundaries. Slow, long-term changes involve alterations in the numbers of cells or in the amount of pigment they contain.

Chromatophores are also present in amphibians and reptiles, but not in birds or mammals, which possess pigment cells called melanocytes. Melanocytes are found mainly in the epidermis, though they occur elsewhere. They also are branched, or dendritic, and their dendrites are used to transfer pigment granules to adjacent epidermal cells. A number of different pigments are produced in the different vertebrate groups, but in mammals only brown eumelanin and yellow or red phaeomelanin are important. Pigment cells—chromatophores and melanocytes alike—are influenced by melanocyte-stimulating hormones of the pituitary.

Epidermal Scales

Epidermal scales are horny, tough extensions of the stratum corneum. Well developed in reptiles, they are also common on exposed skin in birds and mammals. Such scales are periodically molted or shed gradually along with the rest of the stratum corneum. Epidermal scales are absent in fishes, but dermal, or bony, scales are abundant. Clawlike epidermal scales are present in certain amphibians, including a few toads, certain burrowing, wormlike caecilians, and the salamander Hynobius. The so-called horns of the horned lizard are specialized epidermal scales; and the rattle of rattlesnakes is a series of dried scales loosely attached to each other, the last one always remaining despite molting of the rest of the stratum corneum. Epidermal scales cover the bony scales of the carapace (top) and plastron (bottom) of turtles' shells. The beak of turtles is composed of a modified epidermal scale covering the jawbone.

Bills of birds are similarly constructed. In birds, epidermal scales are confined to the lower legs, feet, and base of the bill. The spurs of some birds are bony projections covered with a scalelike sheath. The skin of the webs in aquatic birds is also scaly. In mammals, except for a few cases, epidermal scales are largely restricted to the tails and paws. The overlapping horny plates of the pangolin are modified epidermal scales.

Claws, Nails and Hooves

In many animals, hardened corneal growths occur at the end of the digits, growing parallel to the skin surface. True claws—found in reptiles, birds, and mammals—consist of a dorsal scalelike plate (unguis) covering a ventral plate (subunguis), the whole capping the bony tip of a digit. Nails—found only in mammals—consist of a broad and flattened unguis, with the subunguis reduced to a vestige under the outer tip. Hooves, the characteristic feature of the hoofed mammals, or ungulates, are exaggerated nails, with the unguis curved all around the end of the digit and surrounding the subunguis.

Horns and Antlers

Horns are hardened corneal projections of several types. Except for certain lizards, horns are found only in mammals. The keratin fibre horn is unique to the rhinoceros. It consists of a cone of keratinized cells that grows from an epidermis covering a cluster of dermal bumps (papillae). The fibres, somewhat resembling thick hair, grow from the papillae, and cells between the papillae produce a cement that binds the fibres together.

Hollow horns are found in cattle, sheep, buffalo, goats, and other ruminants. In certain species only the males display them. Such horns consist of an extension of the frontal bone, a permanent part of the cranium covered by a horny layer. The horn of the pronghorn antelope is unique in that the horny covering is shed periodically and a new one is formed from the epidermis that persists over the bony extension.

Antlers, which are characteristic features of the deer family, are not integumentary derivatives at all. Fully developed antlers are solid bone, without any epidermal covering. The young antlers, however, are covered with skin having a velvety appearance. When the antler is fully developed, the drier skin cracks and is rubbed off by the animal. Antlers in giraffes are small and remain permanently covered.

Feathers and Hair

Birds and mammals display remarkable elaborations of the epidermis in the form of feathers and hair, respectively. These distinctive features are dealt with below.

Dermal Derivatives

Dermal scales are found almost exclusively in fishes and some reptiles. They are bony plates that fit closely together or overlap and form the dermal skeleton. Highly developed dermal scales are seen in turtles, where the bony plates form a rigid dermal skeleton that is attached to the true skeleton. In other reptiles, dermal scales are small and localized on parts of the body, as in crocodilians, certain lizards, and a few snakes.

Birds lack dermal scales, and only a single living mammal—the armadillo—displays them. Associated with the evolutionary tendency toward elaboration of epidermal extensions in birds and mammals, there has been a corresponding reduction in dermal derivatives. The membrane bones of the skull, the mandible (lower jaw), and the clavicles (collarbones) are the remaining vestiges of dermal plates in these groups.

Muscular System

The muscular system provides the mechanical activity for the animal in the form of mobility of the different parts of the skeleton or its appendages, the movement of materials along tubular organs such as the alimentary canal, air passages and blood vessels, and the pumping of the blood through the circulatory system by the heart.

Muscles are structured from special muscle cells in the form of fibres that have the ability to contract or shorten. When they relax the muscle lengthens.

Muscle Types

There are three types of muscle found in the animal body. These are:

- Involuntary muscles found in the walls of the alimentary canal, blood vessels, air passages and other tubular structures. These muscles are beyond the control of the will and are called involuntary muscles. The fibres of these muscles do not carry transverse striation or stripes and are therefore said to be 'unstriped' or 'unstriated'.
- Cardiac muscle of the heart. This too is involuntary muscle but is striated and is structured differently to other muscle. It is nucleated, contains many Purkinje Fibres and forms a syncytium with many nuclei but no differentiation of the protoplasm into cells.
- The striated or striped, voluntary muscles of the body that move the various parts of the skeleton or appendages. These consist of very minute thread-like muscle fibres in bundles enclosed by sheaths of fibrous tissue.

Structure of a Skeletal Muscle

The typical voluntary muscle consists of bundles of long muscle fibres. Each fibre is made of long strands called myofibrils that in turn consist of segments called myofilaments. The myofilaments consist of myosin filaments and actin filaments arranged so that the myosin filament is a core surrounded by the actin filaments that are not continuous but form a cap-like structure at each end. Skeletal muscles are attached to the bones by very strong, fibrous bands or cords called tendons.

Muscle Contraction

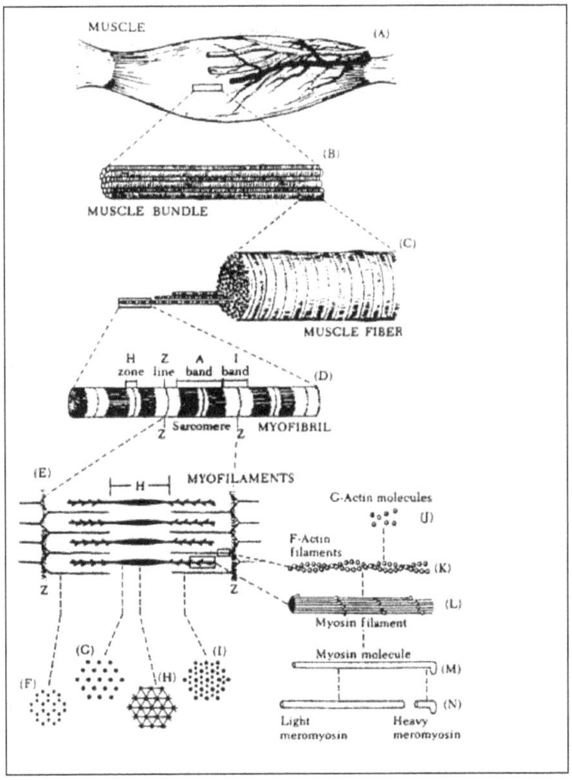

Muscle contraction occurs as a result of a stimulus usually originating from the nervous system either on a voluntary or involuntary basis. When muscle contraction occurs the myofibril segments shorten as a result of the actin filament caps sliding along the myosin filaments. Bands across the myofibril produce the striped effect of skeletal muscle. The muscular system provides the power for the animal and it is here that much of the energy contained in the diet is utilised for normal involuntary and voluntary activity.

Types of Skeletal Muscle Fibres

There are white and red types of skeletal muscle fibres found in birds and all muscles have some white fibres and some red fibres. However, the proportion varies and some muscles are predominantly white and others predominantly red or dark. White fibres lack a compound called myoglobin, but store more glycogen and has a fast contraction of short duration. They have little staying power. The breast muscles of fowls, the muscles of flight, are predominantly white fibres and fowls have very poor flying ability. They fly very short distances with a very rapid wing movement. Red fibres have myoglobin and other cellular structures for continuous production of energy for contraction. These fibres have a slow contraction of long duration. The flight muscles of flying birds consist mainly of red fibres.

Rigor Mortis

When an animal dies the characteristic energy reactions become irreversible and the muscles contract and do not relax for several hours. The stiffening effect of the muscle contraction is called rigor mortis. This condition develops at a faster rate in warmer temperatures or if the animal is fatigued. Relaxation of the muscles usually occurs 10-12 hours after the onset of rigor due to subsequent changes in the body.

Invertebrate Muscle Systems

Cnidarians

The phylum Cnidaria includes the hydras, jellyfishes, and sea anemones. Cnidarians have two main body forms: The cylindrical tentacled polyp, exemplified by the hydra and the sea anemone, and the bell-shaped (or inverted saucer-shaped) medusa. Hydras are some of the simplest multicellular animals to have muscle. They are hollow, cylindrical, freshwater creatures about 10 mm long. One end attaches to a plant or some other support, and the other end is free and has a mouth surrounded by tentacles. The body wall consists of two layers of cells with a middle gelatinous layer called mesoglea. In hydras and other two-layered animals, one kind of cell serves as both muscle and epithelial cells. The compact body of each cell is packed closely with the adjacent cells to form an epithelium, and the base of each cell, where it meets the mesoglea, is drawn out into a long muscle fibre.

In the hydra the musculoepithelial cells that cover the outer surface of the body have longitudinal muscle fibres; those that line the gut cavity (the gastrodermis) have circular muscle fibres. Sea anemones have all of the muscle fibres in the gastrodermis, though some of the fibres are longitudinal and some are circular. When the mouth of the sea anemone is closed, the water in the gut cavity acts as a hydrostatic skeleton, permitting the animal to grow longer and thinner or shorter and fatter or to bend in any direction. These changes result from the interaction of the longitudinal and circular muscles through movements that are not as simple as those in the schematic worm shown in Figure. The hydra can reduce its volume by using its muscles to squeeze water out of the gut cavity through the open mouth. It can reinflate using cilia to circulate water into the gut cavity. Its movements are also influenced by the viscoelastic properties of the mesogleal jelly.

The largest and most familiar medusae are the jellyfishes of the class Scyphozoa, some of which grow to a diameter of two metres. Though large, the scyphozoan jellyfishes have only a single layer of cells on the outer surface of the body and a single layer lining the gut cavity; most of the volume of the animal is occupied by the gelatinous mesoglea. The epidermis of the undersurface of the bell includes the musculoepithelial cells responsible for the animal's weak swimming movements. The muscle fibres contract, reducing the diameter of the bell and forcing out a stream of water. The bell then returns to its original shape by elastic recoil of the mesoglea. These movements are performed in a regular rhythm with a period of a few seconds, propelling the animal through the water. Medusae are among the simplest animals that use muscles to make rhythmic movements. In at least some medusae, the circular muscles, which do most of the work of swimming, are striated. In contrast, most of the other muscles of cnidarians are smooth.

Multilayered Animals

Worms

Although all worms have more than two layers of cells and most have long slender bodies, the various groups of worms are different from each other in other respects.

The simplest worms are the flatworms (phylum Platyhelminthes), most of which have flattened shapes like leaves or ribbons. Although musculoepithelial cells have been found in some flatworms, the muscle cells in most are distinct from the epithelial cells. There is a layer of circular muscle fibres immediately under the epidermis, a layer of diagonal fibres, and a still deeper longitudinal layer. There are also dorsoventral muscle fibres running from the upper to the lower epidermis of the flattened body. These sets of muscle fibres act in various combinations to make the body long and thin, short and fat, or bent to one side or the other. These muscles are also used by some of the larger flatworms to pass waves of muscular contraction along the body, enabling the worm to crawl in a snail-like fashion.

Many flatworms have a mouth opening connected to the pharynx, a muscular tube that carries food from the mouth to the intestine. In some flatworms the pharynx is protruded and inserted into invertebrate prey, to digest and suck out the contents. The sucking is done by peristalsis, waves of muscular contraction that move along the tube from the mouth toward the gut. Although the muscle cells of flatworms are generally not musculoepithelial, their nuclei are found in large cell bodies. The muscle fibres of vertebrates and higher invertebrates, on the other hand, have no projecting cell body.

Roundworms (phylum Nematoda) also have large cell bodies on their muscle cells, but these muscle cells are unique in that nerve fibres do not travel to them as they do in the muscles of other animals. Instead, narrow projections of the muscle cell bodies extend to the principal nerves and contact nerve cells there.

Roundworms have obliquely striated, longitudinal muscle but no circular muscle. They are enclosed in a thick cuticle that allows bending but prevents swelling. Therefore, contraction of the longitudinal muscle can only bend the body. Roundworms do not bend from side to side like eels or snakes, but up or down (dorsal or ventral). By preventing swelling, the cuticle ensures that shortening of one muscle group stretches the other; thus, it makes the dorsal and ventral longitudinal muscles antagonistic to one another. Most crawl between soil particles or among the villi of a host's gut by undulating waves of muscular contraction. Similar movements also enable some roundworms to swim.

The segmented worms (phylum Annelida) include the earthworms and many marine worms. Inside the body, between the body wall and the gut, is a fluid-filled cavity, the coelom, which in some annelids, including earthworms, is divided into successive segments. The body wall has an outer layer of circular muscle and an inner layer of longitudinal muscle.

Earthworms crawl by peristaltic contractions of the body wall. Each segment is alternately elongated (by contraction of its circular muscles) and shortened (by contraction of its longitudinal muscles). The muscles of each segment contract just after those of the segment in front, so that waves of contraction pass backward along the body, enabling the worm to move slowly forward. The same movements also serve for burrowing. While shortened, the segments are pushed against the burrow wall; when they elongate again, the worm moves forward.

Mollusks

The phylum Mollusca includes the gastropods (snails, slugs, and periwinkles), bivalves (clams, oysters, mussels, and scallops), cephalopods (octopods and squids), and other, smaller classes. All mollusks, except the cephalopods, have a highly muscular organ called the foot, through which muscle fibres run in all directions. The foot of a gastropod is a flat structure used for crawling. Waves of muscular contraction travel along its length, moving the animal slowly over the ground. The foot of a bivalve mollusk is a bulbous or tonguelike organ that is used for burrowing in sand or mud. The foot pushes down into the substrate, swells to anchor itself, and then pulls the rest of the animal down behind it.

In addition to the muscles of the foot, gastropod and bivalve mollusks have large muscles attached to their shells. The columellar (shell) muscles of gastropods pull the foot and other parts of the body into the shell. The adductor muscles of bivalves (Figure) shorten to close the shell or relax to allow the shell to spring open, enabling the mollusk to extend its foot or to feed. The adductor muscle can shorten rapidly and close the shell quickly. The muscle is also capable of maintaining the tension needed to hold the shell shut against the spring action of the hinge ligament without using much metabolic energy. Economy of energy is particularly important if the shell has to be kept closed for long periods—for example, for several hours while the mollusk is exposed on the beach at low tide. Fast muscles can shorten rapidly because their cross bridges detach and reattach quickly; however, they use much energy while maintaining tension because there is an energy cost every time a cross bridge detaches and reattaches. Muscles that are economical in their energy usage are generally slow. Accordingly, most bivalve mollusks have two parts to their adductor muscles: a translucent part, which is fast, and an opaque part, which is slow but economical.

Squids and other cephalopod mollusks also swim by jet propulsion. They draw water into the mantle cavity (the cavity that houses the gills) and expel it rapidly. Vigorous movements of this kind provide jet propulsion, but gentler ones serve for breathing by circulating water, and thus oxygen, through the mantle and gills. Fast-swimming squid have mantle cavities whose muscular walls make up as much as 35 percent of the mass of the body.

These walls mainly consist of circular muscle fibres that squeeze water out of the mantle cavity when they contract. Other fibres run radially through the thickness of the wall. These fibres make the wall thinner when they contract, stretching the circular muscle and enlarging the cavity again. Cephalopods do not have longitudinal muscle fibres; however, layers of collagen fibres on the outer and inner surfaces of the muscle prevent the animal from lengthening when the muscles contract. Thus, the circular and radial muscle fibres are antagonistic. Enlargement of the cavity, however, is not solely due to the radial muscle fibres; the cavity tends to expand by elastic recoil of the tissues when the circular muscles relax.

Though many mollusks have shells, most molluscan muscle systems depend on the principle of the hydrostatic skeleton. In some cases, body fluids are involved; for example, the feet of clams are extended and inflated by the inflow of blood. In other cases the muscle itself serves as the incompressible element that must thicken as it shortens or become slender as it elongates, to maintain constant volume. Examples include the shell muscle of the abalone and the tentacles of squid, which are shortened by contraction of longitudinal muscle fibres and lengthened by circular and transverse ones.

Arthropods

Arthropoda is the largest phylum of invertebrate animals and comprises crustaceans, insects, arachnids (spiders and scorpions), and other classes. Some arthropods have soft-bodied young stages in which the principle of the hydrostatic skeleton is important. Most adult arthropods are encased in a skeleton with jointed appendages formed from a stiff cuticle that is divided into separate plates to assist in movement. This skeleton, working as a system of levers, is largely responsible for making muscles antagonistic.

The wing muscles of dragonflies (Odonata) and those of some other insects are worked in simple, direct ways by pulling on the wing bases and making them pivot about their joints. More-advanced insects, including flies (Diptera), work their wings indirectly by muscles that attach to other parts of the skeleton. Although the details of the mechanisms are complicated, the basic principle is simple. Each wing-bearing segment of the body is enclosed by two main plates of cuticle, a tergum above and a sternum below. These plates are flexible enough to be distorted by muscle action. Distortions of the tergum are particularly important in the wing mechanism.

The principal wing muscles are the dorsoventral muscles, which run vertically from the sternum to the tergum, and the longitudinal muscles, which run lengthwise along the segment. Contraction of the longitudinal muscles makes the tergum bow upward, and contraction of the dorsoventral muscles pulls it down again. The wings have joints connecting them to the tergum and to the sternum. Upward movement of the tergum (from contraction of the longitudinal muscles) lowers the wings, and downward movement (from contraction of the dorsoventral muscles) raises them.

All arthropod muscles seem to be striated, not obliquely striated or smooth, and the sarcomeres are of varying lengths. In locusts the sarcomeres of wing muscles are 3.9 micrometres (μm) long, but the sarcomeres of leg muscles (which do not have to contract so quickly) are 8.5 μm long. Wing muscles in many other insects have shorter sarcomeres, often about the same length as those in mammalian muscle (about 2.5 μm).

The force exerted by the muscle is controlled by varying the frequency of action potentials in the axons (an extension of the nerve cell that conducts nerve impulses away from the cell body). The higher the frequency, the larger the force, within limits. In contrast, in vertebrates each muscle is served by many motor axons, each of which is connected to only a small group of muscle fibres. In the twitch muscles that predominate in vertebrates, each muscle fibre is either inactive or fully active, and force is varied by recruiting different numbers of muscle fibres. Like those of other animals, most arthropod muscles require an action potential to initiate each contraction.

Fibrillar muscle is found in the sound-producing, or tymbal, muscles of some cicadas and in the wing muscles of several orders of insects, including the Diptera (flies), Coleoptera (beetles), Hymenoptera (wasps), and Hemiptera (bugs). Most fibrillar muscles work at high frequencies, often of several hundred cycles per second, but they are kept working by action potentials arriving at much lower frequencies. They contract at the resonant frequency of the tymbal or of the wing system. Clipping the wings of an insect that has fibrillar wing muscles increases the frequency of the wing beat, because reduction of the vibrating mass increases the resonant frequency.

All insect wing muscles work aerobically and produce high power outputs. Consequently, they need many mitochondria (the site of aerobic energy production in cells), which may occupy 40 percent or more of their volume in both fibrillar and non-fibrillar muscles. Non-fibrillar muscles that work at high frequencies also need large sarcoplasmic reticulums, but fibrillar ones do not.

Although insect muscles seem to always work aerobically, some crustacean muscles can work anaerobically. The leg muscles that the crab Callinectes uses for swimming include two types of fibres. One type resembles the red muscle fibres of vertebrates in that it is deep pink and contains a high proportion of mitochondria. The other resembles vertebrate white fibres because it is white, with far fewer mitochondria, and presumably works anaerobically. Similar differences occur in other crustacean muscles. Crabs use anaerobic metabolism for short bursts of violent activity in the way that vertebrates do.

Echinoderms

The phylum Echinodermata comprises the starfishes, sea urchins, and their relatives. Their internal skeletons are made of porous blocks of calcium carbonate, and they have muscles to work their skeleton. Echinoderms also have a hydraulic system, the water-vascular system, with movable projections from the body called tube feet.

The details of the tube feet differ among the different groups of echinoderms. In the arrangement found in sea urchins, five double rows of tube feet project through the test, so every part of the body surface has tube feet near it. The tube feet are slender tubes, with a sucker on the closed end. Muscles in the sucker enable it to attach to objects, so the tube feet can be used by the animal to anchor, to move, or to manipulate its prey. Connective tissue in the tube feet limits their diameter but allows them to lengthen, to shorten, and to bend. The tube feet have only longitudinal muscles, which extend the length of the cavity of the tube foot. They are extended by water that has been forced into them by muscles in the wall of the ampulla at their bases.

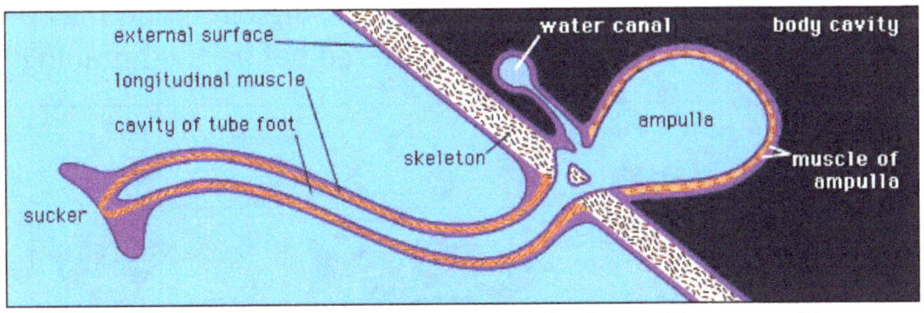

Details of a tube foot of a sea urchin tube foot of the sea urchin.

Vertebrate Muscle Systems

Major Types of Vertebrate Muscles

In terms of its microscopic structure, the musculature of vertebrates is usually divided into three types: striated, cardiac, and smooth muscle. Smooth and cardiac muscle are under the control of the involuntary, or autonomic, nervous system. Striated muscle, on the other hand, is mainly under the control of the voluntary, or central, nervous system. Smooth and cardiac muscle are also similar in their development, being generally associated with the yolk sac. Striated muscle develops directly from the middle of the three embryonic layers, arising largely from the mesodermal somites. In the adult, smooth and cardiac muscle are associated with organs or tubes (viscera), and striated (skeletal) muscle is associated with the bony or cartilaginous skeleton.

The two major divisions of the vertebrate musculature are the visceral musculature and the somatic musculature (the striated muscles of the body wall). Somatic musculature may be divided into appendicular, or limb, muscles and axial muscles. The axial muscles include the muscles of the tail, trunk, and eyeballs as well as a group of muscles called hypobranchial muscles, which separate and migrate from the others during development.

Basic Pattern of Development

The gastrula is the stage of embryonic development at which the embryo appears as three distinct layers of cells (the germ layers): The exterior ectoderm, the middle mesoderm, and the interior endoderm. The mesoderm differentiates to form most of the tissues, structures, and organs of the body. As the embryo lengthens, the mesoderm lying along the midline differentiates to form the notochord, a hollow cartilaginous nerve tube. In the adult the notochord contributes only to the structure of the vertebrae. The mesoderm lateral to this midline then divides into three parts that ultimately form the somites (which subsequently form the vertebrae, the somatic muscles, and the skin), the structures of the urogenital system, and the coelom and its associated structures.

Comparative Anatomy

One problem in discussing the differences in arrangement of muscles between the various vertebrate groups is in deciding which muscles in each species are homologous—that is, which have the same evolutionary and developmental origin. The problem arises because the position and attachment of muscles change during evolution; a muscle lying in the same position and attached to the same bone or cartilage in one vertebrate may have different origins from those of another vertebrate species. Comparison of the development of muscles in the embryo of each species and of their nerve supply would probably give the best clues. No single method may be relied upon in all cases, and many different types of evidence are considered before the homology is decided upon.

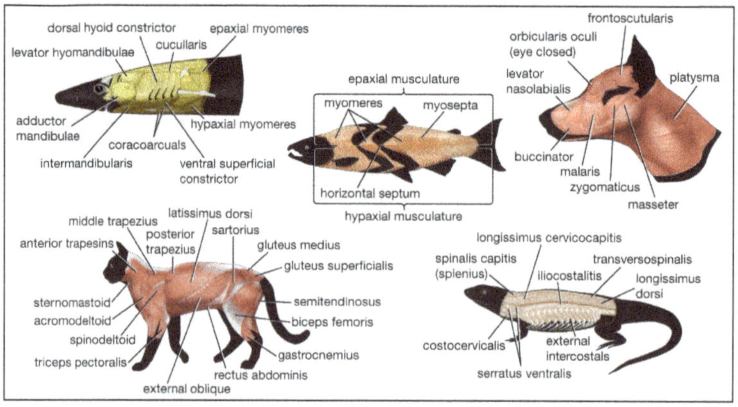

Vertebrate musculature: Trunk musculature of five representative vertebrates.

Vertebrate muscles are given names derived from Latin according to their attachments. In this system the Latin names of the bony points of attachment are either joined, as in sternocleidomastoid, naming the human muscle that runs from the sternum and clavicle to the mastoid region of the skull, or they may be named for their form or their gross function. There are several standard terms that describe form and function. A muscle may have more than one point of origin; thus, it may be described as having, for example, two "heads," as in biceps femoris (bi- for two, -ceps for heads, femoris meaning "of the femur"). It may be long, longus, or short, brevis. It may run transversely across a body segment, transversus, or obliquely, obliquus. It may lie close to the surface, superficialis, or deep, profundus. In describing function, flexors are muscles that tend to close the angle made by the two bones to which they are attached; extensors tend to increase the angle. Adductors pull a bone or cartilage closer to the axis of the body, or limb, while abductors pull away from the axis. Rotators turn one bone or cartilage with respect to another or with respect to the midline. Pronators turn the sole of the foot or the palm of the hand to face the ground, while the opposite function is performed by supinators. Constrictors and sphincters diminish the volume of spaces or the area of structures, and dilators increase them. The names of muscles in humans often have been applied to grossly equivalent muscles in animals, a situation that often causes confusion.

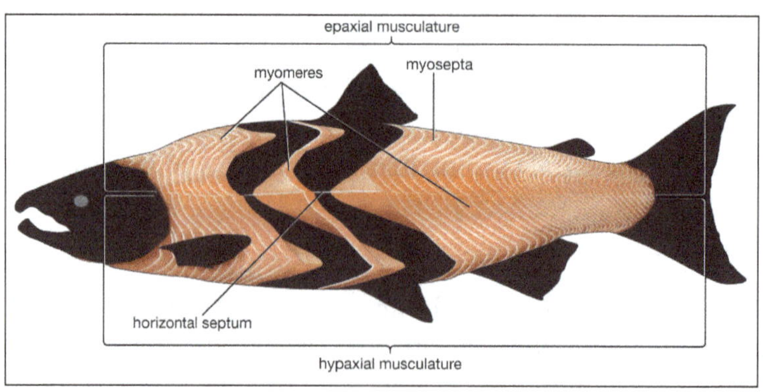

Trunk musculature: Salmon Lateral view of the trunk muscles of a salmon.

Jawless Fishes

The earliest known vertebrates were jawless fishes of the class Agnatha, and their only living representatives are the cyclostomes—the lampreys and the hagfishes. The modern agnathans retain much of the general organization of the ancestral vertebrates, and, therefore, much of their musculature is relevant to an understanding of the evolution of muscles in more-advanced vertebrates.

The cyclostomes are free-swimming animals with prominent axial somatic musculature, which during contraction produces undulating waves that propagate from head to tail to produce thrust. The axial muscles form a single segmented mass running vertically down the side of the fish. These muscle segments, known as myomeres, consist of relatively short fibres that insert into septa of connective tissue, the myocommata, between the adjacent myomeres. There is only a rudimentary axial skeleton and no appendicular skeleton, so there are no limb muscles. The eyes of cyclostomes are degenerate structures, and the six axially derived muscles normally found associated with vertebrate eyes are diminished or absent. The branchiomeric muscles in cyclostomes are represented by a sheet of constrictors that compresses the gill pouches and helps the pumping mechanism draw water through the pharynx to the gills. Other muscles of the branchiomeric series have been modified for specialized feeding functions. The branchiomeric musculature of more primitive jawless fishes would probably have been similar for each of the gill arches.

Jawed Fishes

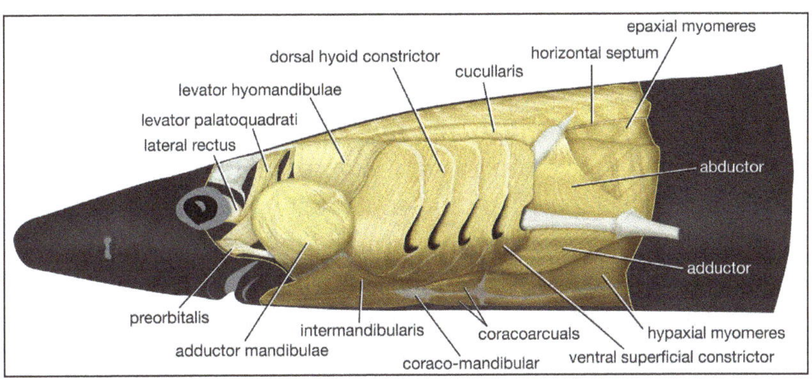

Anterior musculature: Dogfish Lateral ventral view of the anterior muscles of a dogfish.

The sharks and other cartilaginous fishes (the class Chondrichthyes) have modified the structure of the first two arches; the cartilages of the anterior arch form the mandible and upper jaw (palatoquadrate), and modifications also have taken place in the second, hyoid arch. The posterior five gill arches of more primitive sharks, however, are a good model for the condition in the ancestral jawless fishes. Each arch has a visceral skeleton comprising five cartilages named, from dorsal to ventral, the pharyngobranchial, epibranchial, ceratobranchial, hypobranchial, and basibranchial. The cartilages are arranged at angles to each other. Each cartilaginous arch is provided with a set of branchial muscles that receives separate, visceral innervation. Superficially, a thin sheet comprising dorsal and ventral constrictor muscles runs in the flap of skin that covers each gill slit and forms the gill septum. Most fibres attach, dorsally and ventrally, to connective tissues (fascia) that sheath the body. Some of the deeper fibres attach to the gill bar and may run between adjacent bars. These thin, broad muscles squeeze the pharynx closed as part of the pumping action necessary for gill breathing. Dorsal and deep to this layer, a levator muscle runs from the sheathing fascia to the pharyngobranchial, and it can elevate the gill arch. In some sharks, however, the most posterior sets of levator muscles, whose fibres run diagonally down and back, may join adjacent levators, become enlarged, and attach to the pectoral girdle. This mass is known as the trapezius and evolves into the tetrapod muscle of the same name. Adductor muscles are positioned so as to close the angle between the epibranchial and ceratobranchial, and an interarcual muscle performs the same function for the angle between the pharyngobranchial and epibranchial cartilages.

In the jawed fishes, including the sharks, the axial musculature of the trunk and tail (a single block in cyclostomes) differentiates into dorsal and ventral components, which are separated by connective tissue. The dorsal block of muscle is known as the epaxial musculature, and the ventral block, the hypaxial. The epaxial block runs from the back of the skull to the end of the tail, while the hypaxial block is not present any farther forward than the pectoral (shoulder) girdle (because of the presence of the branchial [gill] apparatus). The hypaxial musculature in the tail forms a solid block of muscle, while in the trunk it encloses the body cavity. Ribs develop in the horizontal septum separating the two blocks of muscle and usually lie in the myocommata, the fascial tissue separating each myomere. In fishes, the ribs primarily serve to improve the leverage of the myomeres in producing the undulatory movements of swimming. The ribs are short in sharks but may develop to considerable length in bony fishes. Unlike the cyclostomes, where the myomeres form a series of essentially vertical strips of muscle, the myomeres of all jawed fishes are folded in a complex fashion. This development is related to the development of a more powerful swimming ability in the jawed fishes. The myomeres are folded in a zigzag pattern, projecting strongly forward halfway down the side of the fish, with a smaller, backward projection both dorsal and ventral to this point; the effect is of a W on its side. These projections become sharper and more cone-shaped deep to the surface of the fish and thus come both to be overlapped by the folds of several anterior myomeres and to overlap those of several more posterior myomeres. The folding and overlapping of myomeres has the effect that contraction of a single myomere produces curvature over a considerable part of the body of the fish. The fishes that swim faster thus tend to have a greater degree of folding and overlapping. In the tunny, for example, one myomere may have an overlap with 20 others. The undulations of the body and caudal (tail) fin produced by these axial muscles can produce much greater thrust than is produced by the beating of the appendicular fins. The latter are mostly used in slow "precision" swimming, as when a fish is investigating food, while undulations of the body are used for faster, powerful swimming. The axial musculature of fishes contributes up to half the weight of the fish, while the appendicular muscle contributes less than a fifth of the fish's mass.

In all higher vertebrates, the most anterior element in the axial musculature is the set of six eye muscles derived from the three pre-otic somites (those anterior to the ear region of the embryo). The rectus muscles move the eyes about the longitudinal axis of the body, that is, superiorly (upward) or inferiorly (downward), or about a vertical axis, in other words, laterally (backward) or medially (forward), according to their position relative to the eyeball. The oblique muscles, superior and inferior, rotate the eyes about a transverse axis.

Jawed fishes have single midline fins and two sets of paired fins. The unpaired dorsal and anal fins of teleosts (advanced bony fishes) have axially derived muscle sheets on either side, which, when contracted, may change their angle and even fold the fins. The paired pectoral fins and the weaker pair of pelvic fins, however, have a mass of musculature both dorsal and ventral to them that is derived from mesenchymal cells. The dorsal muscle mass lifts the fin or pulls it posteriorly; the ventral mass pulls it down or forward. The two major muscle masses are attached at one end to the pectoral or pelvic girdle and on the other to the base of the fin. The amount of downward or upward movement of the fin versus the amount of backward or forward movement can be adjusted, in some fishes, by small slips of muscle derived from the major dorsal and ventral masses, which twist the fin.

The hypobranchial muscles of jawed fishes are straplike muscles running from the pectoral girdle to the structures of the visceral skeleton, the jaws, and the gill bars. Some muscles, such as the coracomandibularis, are specialized as jaw openers, although most of the work of jaw opening is done by gravity.

In bony fishes the gill septum of the hyoid arch is greatly modified to become a single, movable, bony covering for the whole gill chamber—the operculum. The individual gill septa are lost, and there is a great modification of the posterior branchial muscles, with many of the elements found in sharks (e.g., levators and adductors) becoming reduced or absent. The superficial constrictor of the hyoid arch in sharks is remodeled in bony fishes to control the opening and closing of this protective cover.

Electric organs appear to have arisen independently in several fishes. They are modifications of the axial musculature of the tail, as in the electric eel Gymnotus, a teleost, or of the muscles of the pectoral fins, as in the ray Torpedo. In a few cases electric organs lie superficially to the musculature and may be derived from modified glandular tissue, as in the Nile catfish Malapterurus.

Origins of the Tetrapod Limbs

The invasion of land led to a complete change in emphasis in the propulsive elements of the muscular system. In fish the axial musculature is much more important as a mover of the body than is the appendicular musculature. The evolution of land vertebrates is characterized by an increasing emphasis on the limbs for propulsion and by a corresponding de-emphasis on the axial musculature. The limbs of tetrapods are generally similar in overall pattern. Primitively at least, most major groups have similar characteristic features: the fore and hind feet have five digits; there is one bone in the proximal part of the limb (nearest to the body) and two in the distal part (away from the body); and there are a wrist or ankle joint, an elbow or knee joint, and a shoulder or hip joint. Although most muscles have several roles, the major actions of tetrapod limb muscles are similar: some primarily resist the downward force of the body at hip and shoulder, others press the supporting fore or hind feet down onto the ground at wrist or ankle or pull back on the supporting limbs (at all three joints) to create thrust, and others primarily pull the "swing" limbs forward into a new support position.

The limbs may originally have developed more as supportive struts. Structurally, the tetrapod limb can be derived from the pattern found in the paired fins of Sarcopterygii, a class of lobe-finned fishes. These were once a large radiation but have been largely replaced by the Actinopterygii, the class of ray-finned fishes. Today the lobe-finned fishes are represented by the coelacanth (Latimeria) and the lungfishes (Dipnoi). The lungfishes, denizens of shallow and seasonal waters, habitually use their fins as supports, but propulsion is largely achieved by undulations of the body, as is the case with other fish.

Tetrapod Musculature

In the living urodeles (newts and salamanders) of the class Amphibia, the axial muscles are most important for propulsion. The limbs of urodeles are quite weak and tend to be carried forward passively with the undulations of the body. As the primary propulsive force is provided by the muscles of the trunk, urodeles retain large axial muscles. The axial muscles are still segmented, separated by myocommata, although the myomeres run vertically and without the elaborate folding seen in jawed fishes. The epaxial muscles, given the name dorsalis trunci in tetrapods, are little changed, although some modification has taken place to promote a facility for dorsoventral bending of the spine that occurs in tetrapods but rarely in fishes.

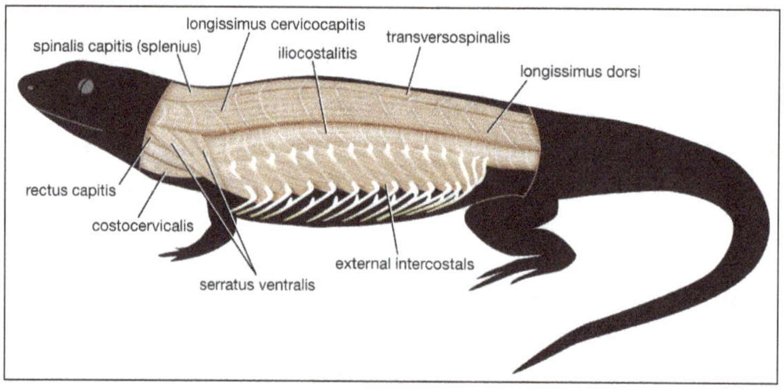

Trunk musculature: Lizard axial view of the trunk muscles of a lizard.

The anurans (frogs and toads) have rather similar, but considerably reduced, epaxial muscles. There is, however, a trend in tetrapods toward finer control of muscular action with increasing complexity. In reptiles the epaxial muscles, although still retaining a semi-segmental structure, are divided into several structural and functional units. The deepest set of muscles, the transversospinalis group, are short and run obliquely forward, over one to four vertebrae, from the transverse process of one vertebra to the lamina (the flat plate of bone at the base of the vertebral spine) of a more anterior vertebra. The transversospinalis group is particularly responsible for rotatory movements of the spine. Superficial to transversospinalis lays longissimus, with much longer fibres, which is important in extension of the back. More superficial still and lateral to these muscle blocks is iliocostalis, a flat sheetlike muscle that runs from the pelvic girdle upward and laterally to attach to the ribs. It is particularly important in lateral flexion (bending) of the spine. This general pattern is further complicated in snakes, which have secondarily returned to the propulsive use of the axial muscles. In birds the vertebral column of the trunk region undergoes much fusion, and this complexity is reduced, as indeed it is in chelonians (turtles and tortoises). Mammals retain the broad pattern of the reptile epaxial musculature but (with the exception of the innervation of the musculature) have greatly reduced the segmentation that is present in reptiles.

In the tails of urodeles the hypaxial muscles are also largely unchanged. As with all land vertebrates, however, the demands of supporting the viscera when living in an air environment have brought about major modifications of the hypaxial musculature of the trunk. In typical tetrapods a strong series of ribs has developed for the same reason. Although urodeles have secondarily reduced their ribs, they show many of the typically tetrapod features of the hypaxial musculature. The muscles fall into three groups. A group of subvertebral muscles forms ventral to the vertebrae, in the region of their joints with the ribs at the transverse processes. It acts in ventral and lateral flexion (bending) of the spine. A rectus abdominis muscle runs longitudinally along the ventral aspect of the body wall between the pectoral and pelvic girdles, and laterally this muscle is associated with the third group, the lateral hypaxial muscles. The third group consists of three major layers of muscle, whose fibres are oriented in differing directions, a feature that gives additional strength to the body wall. Superficially lies the external oblique muscle, with fibres running longitudinally but somewhat ventrally; deep to this lies the internal oblique, with fibres running longitudinally and somewhat dorsally; and deepest lies the transversus muscle, whose fibres run dorsoventrally.

In the higher tetrapods the external and internal obliques tend to become further divided into layers in the abdominal region. The thoracic representatives of these muscles tend to become divided into discontinuous, rather thin muscle layers between the ribs (external and internal intercostals), superficial to the ribs (supercostals), and deep to the ribs (subcostals). While only the rectus abdominis tends to retain visible evidence of segmental origin, in its tendinous intersections (which are present even in humans) the segmental innervation of the hypaxial muscles is retained in all tetrapods.

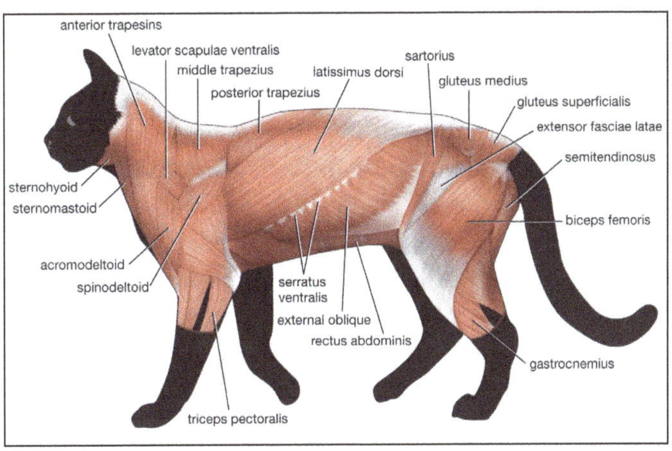

Trunk musculature: Cat Lateral view of the trunk musculature of a cat.

In tetrapods, unlike fishes, the pectoral girdle does not have a solid bony connection to the axial skeleton but rather is supported by a series of muscles derived from the outer layer of hypaxial trunk muscles. This is no doubt another adaptation to life in an air environment, where the cushioning effect of water has been lost. These muscular slings are not readily demonstrated in the living amphibians, which are either skeletally degenerate, as in urodeles, or highly specialized toward leaping, as in anurans (frogs and toads). In more typical tetrapods, there are two major derivatives of the external oblique attaching the scapula (shoulder blade) to the body: the serratus, made up of numerous fingerlike slips running from the scapula to the neighbouring ribs, and the levator scapulae, which are fused with serratus along its caudal (tail-end) border. Levator scapulae consist of fibres running more anteriorly to ribs or transverse processes of the neck. Mammals and some reptiles have a third such muscle, attaching the pectoral girdle to the region of the spine, called rhomboideus. Mammals also have utilized part of the hypaxial musculature to form a muscular septum between the region of the lungs and heart (the thoracic cavity) and the region of the digestive and reproductive viscera (the abdominal cavity). This is the diaphragm, which is the most important respiratory muscle in the mammalian body.

The six axially derived eye muscles of fishes undergo only small modifications in tetrapods. Eye movements are changed, partly according to changes in the orientation of the orbit, such as the trend toward orbital frontality that is typical in primates. Additional eye muscles may be derived by splitting some of these six muscles. An example of this is the retractor bulbi muscle, which is derived from the lateral rectus muscle. In amphibians and some reptiles it pulls the eyeball deeper into the orbit for protection, and in amphibians it is an aid in swallowing. Another example is the levator palpebrae superioris, derived from the superior rectus, which elevates the upper eyelid to open the eye.

The limb muscles of typical tetrapods are derived from the dorsal and ventral muscle blocks of the paired fins of fishes. In tetrapod development, this pattern of derivation from dorsal and ventral muscle blocks is repeated. As a consequence, the homologies of the muscles of the typical tetrapod limb often can be traced by considering the source of innervation of each muscle from the nerves of the dorsal (or extensor) compartment or the ventral (or flexor) compartment.

In the pectoral limb the dorsal, extensor group of muscles includes several that appear consistently and with similar roles. Beginning with the muscles that act on the humerus (the proximal bone of the limb), all tetrapods have a large sheetlike muscle known as the latissimus dorsi that runs from the side of the trunk to the humerus. The latissimus dorsi muscle retracts the humerus and thus propels the body forward. Acting to rotate, flex, or adduct the humerus, depending on limb posture, is a muscle known as subcoracoscapularis in amphibians, reptiles, and birds and as subscapularis in mammals. It runs from the deep surface of the shoulder girdle to the humerus. In amphibians the dorsalis scapulae arise from the anterior edge of the scapula. The same muscle is known as the deltoideus in reptiles and mammals; in the latter, part of its origin moves from the scapula to the clavicle (collarbone). It is a major abductor of the shoulder in most tetrapods. At the elbow joint, all tetrapods have a muscle called triceps as the major extensor. It arises in several heads from the shoulder girdle and humerus. There are always a variable number of extensor muscles for the wrist and digits (fingers and toes) arising from the region of the elbow joint, on the lateral aspect of the humerus.

On the ventral, flexor aspect of the pectoral limb, the pectoralis is found in all tetrapods. The pectoralis runs from the chest wall to the humerus, on which it acts to pull the humerus downward and backward. This muscle not only is important in providing forward thrust in quadrupedal locomotion but is the chief depressor of the forelimb in birds and bats. The major elevator of the wing in birds, supracoracoideus, is present in all tetrapods. In mammals the supracoracoideus retains its attachment to the humerus, but its previous point of origin (the coracoid plate) disappears, and the muscle now appears as two separate blocks of muscle arising on either side of the spine of the scapula as an abductor muscle (supraspinatus) and a rotator and flexor (infraspinatus). Coracobrachialis and (except in amphibians) biceps arise from the tip of the coracoid and act to flex the elbow. In this they are aided by the brachialis muscle, which arises from the humerus. As on the extensor aspect, there are always a number of flexors of the wrist and digits. These arise on the medial side of the distal humerus.

The muscles of the pelvic limb cannot be readily compared beyond the reptiles and mammals. Even in these cases, changes in limb posture have led to major changes in the arrangement and function of muscles. On the dorsal aspect, a single large muscle in reptiles, puboischiofemoralis, runs from the bones of the pelvis to the femur (the proximal bone of the hind limb). This reptilian muscle appears to be represented by three mammalian hip muscles: psoas, iliacus, and pectineus. Iliofemoralis acts as an abductor of the hip in reptiles and appears to be represented by the gluteal muscles in mammals, but the function of the gluteal muscles is different. More similar in reptiles and mammals is the quadriceps or quadratus femoris, which consists of multiple heads (four in mammals) that arise from the pelvic girdle and femur and insert by a common tendon into the tibia (the larger bone of the distal pectoral limb). It is the sole extensor of the knee joint in both the reptiles and the mammals. The extensors of the ankle and digits in both reptiles and mammals are not dissimilar to those of the pectoral limb and take origin from the lateral and anterior surfaces of the two distal bones of the pelvic limb. On the ventral aspect of the hind limb, small, deep muscles run from the internal and external pelvis to the head of the femur and help in adduction and rotation. Of these, the puboischiofemoralis externus of reptiles appears to be represented by the obturator externus of mammals, and, similarly, the ischiotrochantericus of reptiles appears to be the homologue of the obturator internus of mammals. Again, the major adductor of the hip of reptiles, adductor femoris, appears to be homologous with some of the muscles called the adductors in mammals. There seem to be some homologies between the major flexors of the hip and thigh in reptiles, such as puboischiotibialis, and two deeper muscles, flexor tibialis externus and internus, and some functionally equivalent muscles in mammals: the gracilis, semimembranosus, and semitendinosus. In reptiles the axial muscle of the tail is strong, and the caudofemoralis, a powerful flexor of the thigh that originates in the tail, is consequently large. The tail in mammals, although usually present, is much more gracile, and, as a result, caudofemoralis is represented by only a few small muscles. Another major change is in the flexors of ankle and digits. In reptiles, these insert by long tendons passing below the ankle joint, much as in the forelimb. In mammals, however, the equivalent long flexor, gastrocnemius, inserts on a new bony process, the calcaneal tuberosity, or heel bone, which gives more efficient leverage.

The hypobranchial muscles of tetrapods are both reduced and modified in comparison with those of jawed fishes. In tetrapods these straplike muscles still arise from elements of the pectoral girdle but now pass to the new derivatives of the gill arches of fishes: the hyoid bone and laryngeal cartilages. They act primarily in the gross movements of these structures in swallowing and the production of sound—for example, as depressors of the hyoid (sternohyoid, omohyoid) or of the larynx (sternothyroid). Fibres of the hypobranchial muscles in the region of the hyoid are utilized to form the internal musculature of the tongue.

The branchial musculature is also modified in tetrapods from the condition seen in jawed fishes. The development of a shoulder muscle, the trapezius, from the levator muscles of the gill arches of fishes, is taken further in tetrapods by the separation of further slips of muscle to form muscles such as sternocleidomastoid, a muscle important for humans in movements of the head and in breathing. In mammals that lose the clavicle, these slips may be further modified to form muscles running from the head to the pectoral limb. Tetrapods, with the exception of mammals, utilize part of the constrictor muscle of the hyoid arch to form the depressor mandibulae, which replaces the hypobranchial muscles as the major jaw-opening muscle. The restructuring of the posterior jaw in mammal's leads to the further replacement of this new muscle by the digastric, which is a compound muscle made up of parts of the constrictors of the first and second branchial arches. Thus, it is partly innervated by the mandibular division of the fifth cranial nerve (as is the case with other jaw muscles and the tensor tympani, one of the muscles of the ear) and partly by the seventh cranial nerve, the facial nerve (which also supplies an ear muscle associated with the stapes, an ear bone derived from the hyoid arch). The levator palatoquadrati, which elevates the upper jaw in jawed fishes, is retained as a jaw muscle in birds and in some reptiles, as they share the ability of fishes to move the upper jaw. The adductor mandibulae is much altered in tetrapods, although its overall function is retained. During the course of tetrapod evolution, it becomes a superficial muscle, and in mammals it splits into several functional units arising from the undersurface and side of the skull and attaching to various points on the mandible. These are the lateral pterygoid, which pulls the jaw forward; the medial pterygoid and its partner, the masseter, which close the jaw and move it from side to side; and the temporalis, which closes the jaw and pulls it backward. All are innervated by the first-arch cranial nerve, the fifth nerve. The intermandibularis of jawed fishes is retained as the mylohyoid of tetrapods, which is an elevator of the tongue.

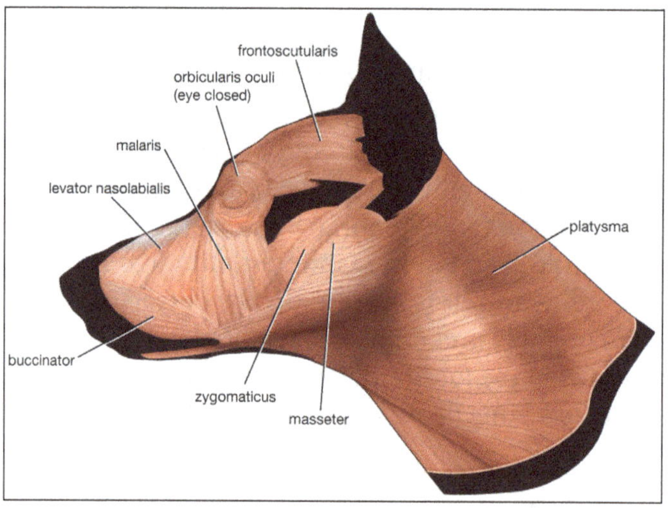

Facial musculature: Dog Lateral view of facial musculature of a dog.

Finally, the constrictor muscle of the hyoid arch, which in bony fishes is used to control the operculum, is remodeled in tetrapods as a sheathing superficial muscle of the neck, the sphincter colli. It derives its innervation from the nerve of the hyoid arch, the seventh, or facial, nerve. This cranial nerve is named from the further adaptation of the sphincter colli muscle in mammals, particularly in higher primates, as the many small muscles of facial expression, which allow people to smile, laugh, and frown.

Skeletal System

The skeletal system is the 'framework' upon which the body is built – it provides support, protection and enables the animal to move. The joints are considered to be an integral part of the skeleton. The skeletal system is made of the specialised connective tissues, bone and cartilage.

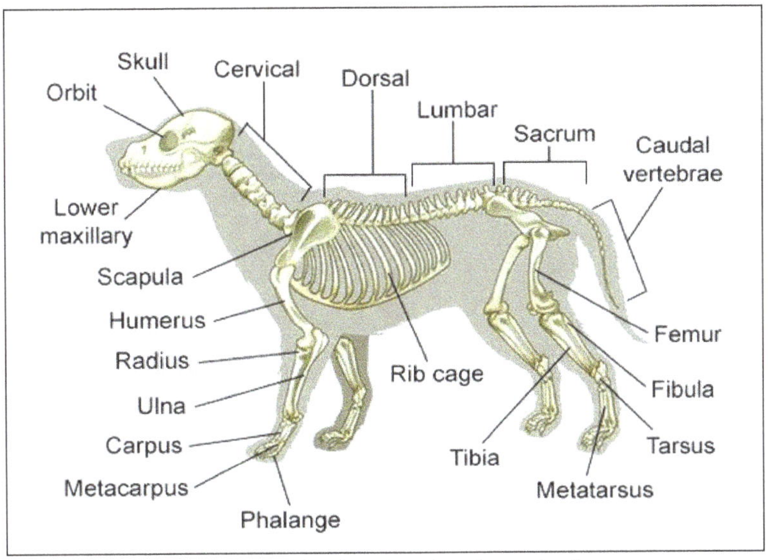

The skeleton of the dog showing the major bones.

The functions of the skeletal system are:

- Support – It acts as an internal 'scaffold' upon which the body is built.
- Locomotion – It provides attachment for muscles, which operate a system of levers, i.e. the bones, to bring about movement.
- Protection – It protects the underlying soft parts of the body, e.g. the brain is encased in the protective bony cranium of the skull.
- Storage – It acts as a store for the essential minerals calcium and phosphate.
- Haemopoiesis – Haemopoietic tissue forming the bone marrow manufactures the blood cells.

Bone Structure and Function

Bone Shape

Bones can be categorised according to their shape:

- Long bones – These are typical of the limb bones, e.g. femur, humerus, and also include bones of the metacarpus/metatarsus and phalanges; long bones have a shaft containing a medullary cavity filled with bone marrow.
- Flat bones – These have an outer layer of compact bone with a layer of cancellous or spongy bone inside; there is no medullary cavity, e.g. flat bones of the skull, scapula and ribs.
- Short bones – These have an outer layer of compact bone with a core of cancellous bone and no medullary cavity, e.g. carpal and tarsal bones.
- Irregular bones – These have a similar structure to short bones but a less uniform shape; they lie in the midline and are unpaired, e.g. vertebrae.

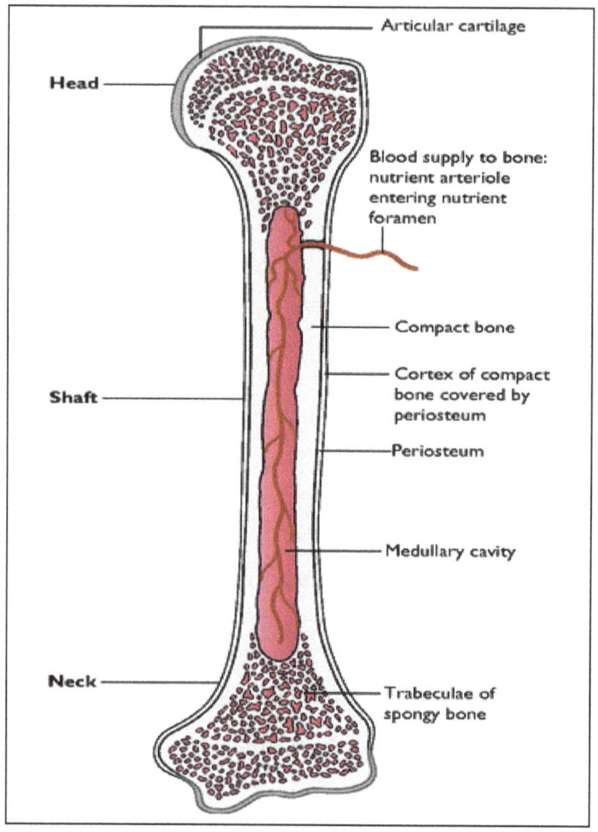

Structure of a typical long bone.

Some specialised types of bone are:

- Sesamoid bones – These are sesame-seed-shaped bones that develop within a tendon (and occasionally a ligament) that runs over an underlying bony prominence; they serve to change the angle at which the tendon passes over the bone and thus reduce 'wear and tear', e.g. the patella associated with the stifle joint.
- Pneumatic bones – These contain air filled spaces known as sinuses that have the effect of reducing the weight of the bone, e.g. maxillary and frontal bones.
- Splanchnic bone – This is bone that develops in a soft organ and is unattached to the rest of the skeleton, e.g. the os penis (the bone within the penis of the dog and cat).

Development of Bone

The process by which bone is formed is called ossification and there are two types: intramembranous and endochondral ossification. The cells responsible for laying down new bone are called osteoblasts; the cells that destroy or remodel bone are called osteoclasts.

Endochondral Ossification

The process of endochondral ossification is as follows:

- A cartilage model develops within the embryo.

- Primary centres of ossification appear in the diaphysis or shaft of the bone. The cartilage is replaced as the osteoblasts lay down bone, which gradually extends towards the ends of the bone.
- Secondary centres of ossification appear in the epiphyses or ends of the bone, continuing the bone development.
- Osteoclasts then start to remove bone from the centre of the diaphysis to form the marrow cavity, while the osteoblasts continue to lay down bone in the outer edges.
- Between the diaphysis and epiphyses a narrow band of cartilage persists. This is the growth plate or epiphyseal plate, which allows the bone to lengthen while the animal is growing. Eventually, when the animal has reached its final size, this will be replaced by bone and growth will no longer be possible. The epiphyseal plate is then said to have 'closed' and the time at which it happens is different for each type of bone.

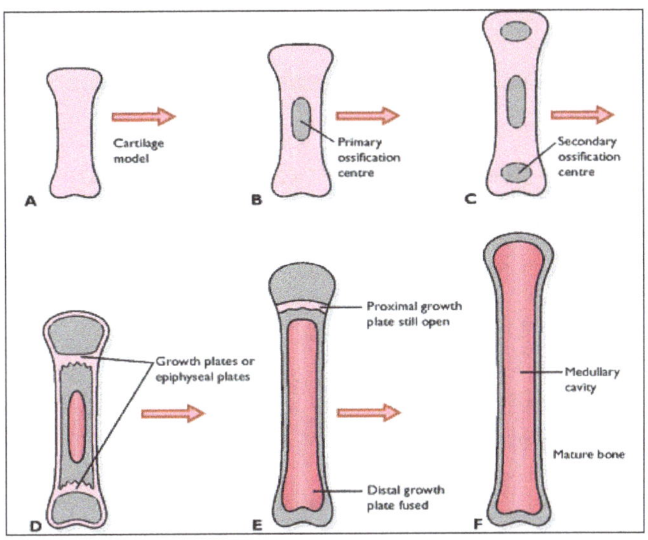

Stages of endochondral ossification.

A cartilage model of the bone exists in the embryo. B Ossification begins from the primary ossification centre in the shaft (diaphysis). C Secondary ossification centres appear in the ends of the bone (epiphyses). D Ossification continues in primary and secondary centres; osteoclasts start to break down bone in the shaft, creating the marrow cavity. The first growth plate fuses and growth is now only possible at the proximal growth plate; the medullary cavity extends into the epiphysis. F The proximal growth plate fuses and bone growth ceases.

The Skeleton

The skeleton can be divided into three parts:

- Axial skeleton – Runs from the skull to the tip of the tail and includes the skull, mandible, vertebrae and also the sternum.
- Appendicular skeleton – The pectoral (front) and pelvic (hind) limbs and the shoulder and pelvic girdles that attach (or append) them to the body.
- Splanchnic skeleton – In the dog and cat, this is represented by the os penis within the tissue of the penis.

When studying the anatomy of the skeletal system it is helpful to understand the terms that are used to describe the various projections, passages and depressions that are found on and within bones.

The Axial Skeleton

The Skull

The bones of the head include the skull, nasal chambers, mandible or lower jaw and hyoid apparatus. The functions of the skull are:

- To house and protect the brain.
- To house the special sense organs – eye, ear, nose and tongue.
- To house and provide attachment for parts of the digestive system – teeth and tongue, etc.
- To provide attachment for the hyoid apparatus and the numerous muscles of mastication and facial expression.
- To provide a bony cavity through which air can enter the body.
- To communicate – the muscles of facial expression are found on the head and are an important means of communication.

Cranium

The caudal part of the skull that provides the bony 'case' in which the brain sits is called the cranium. The bones of the cranium include:

- Parietal – Forms much of the dorsal and lateral walls of the cranium.
- Temporal – Lies below the parietal bone on the caudolateral surface of the skull. The most ventral part of the temporal bone forms a rounded prominence called the tympanic bulla, which houses the structures of the middle ear. There is an opening into the tympanic bulla, called the external acoustic meatus, which in life is closed by the tympanic membrane or eardrum. The cartilages of the external ear canal are attached to this region.
- Frontal – Forms the front aspect of the cranium or 'forehead'. Contains an air-filled chamber called the frontal sinus, which connects to the nasal chamber.
- Occipital – This lies at the base of the skull on the caudal aspect. In this region there is a large hole called the foramen magnum, through which the spinal cord passes. On either side is a pair of bony prominences, the occipital condyles. These articulate with the first cervical vertebra or atlas. At the side of the occipital condyles are the jugular processes, which are sites for muscle attachment.
- Sphenoid – This lies on the ventral aspect of the skull, forming the floor of the cranial cavity. It is penetrated by many small foramina through which nerves and blood vessels pass.
- Sagittal crest – A ridge of bone on the dorsal midline surface of the skull, which can be prominent in muscular dogs.

- Zygomatic – The zygomatic arch is an arch of bone that projects laterally from the skull, forming the 'cheekbone'.
- Lacrimal – Lies at the base of the orbit, which houses the eye, and is the region through which the tears drain from the eye into the nose.

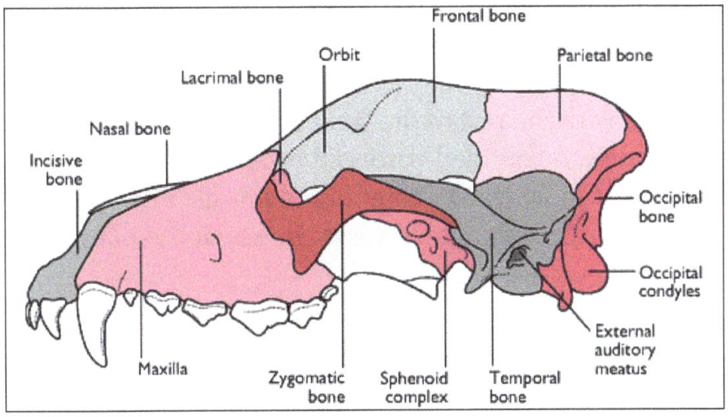

Lateral view of the dog skull showing main bony features.

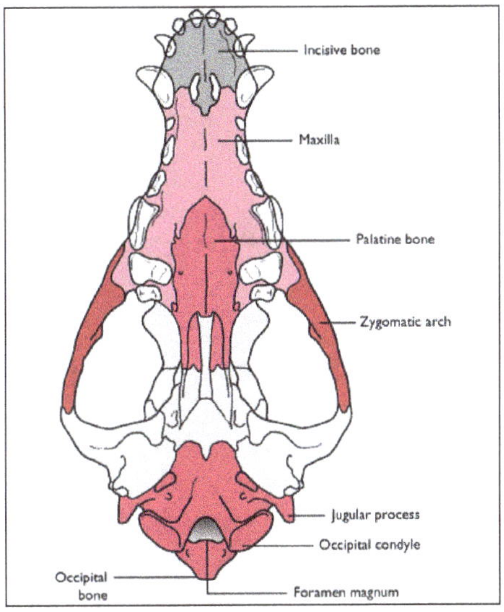

Ventral view of the dog skull showing main bony features.

Nasal Chambers

The most rostral part of the skull carries the nasal chamber, the sides of which are formed by the maxilla and the roof by the nasal bone. The nasal chamber is divided lengthways into two by a cartilaginous plate called the nasal septum. Each of the chambers is filled with delicate scrolls of bone called the nasal turbinates or conchae. These are covered in ciliated mucous epithelium. At the back of the nasal chamber, forming a boundary between the nasal and cranial cavities is the ethmoid bone. In the centre of this bone is the cribriform plate – a sieve-like area perforated by numerous foramina through which the olfactory nerves pass from the nasal mucosa to the olfactory bulbs of the brain.

The roof of the mouth is called the hard palate and is formed from three bones on the ventral aspect of the skull: The incisive bone or premaxilla, part of the maxilla and the palatine. The incisive bone is the most rostral and carries the incisor teeth.

Many of the bones of the skull are joined together by fibrous joints called sutures. Sutures are firm and immovable joints but allow for expansion of the skull in a growing animal.

Mandible

The mandible or lower jaw is comprised of two halves or dentaries, joined together at the chin by a cartilaginous joint called the mandibular symphysis. Each half is divided into a horizontal part, the body, and a vertical part, the ramus. The body carries the sockets or alveoli for the teeth of the lower jaw. The ramus articulates with the rest of the skull at the temporomandibular joint via a projection called the condylar process. A rounded coronoid process, which projects from the ramus into the temporal fossa, is the point to which the temporalis muscle attaches. There is a depression on the lateral surface of the ramus, the masseteric fossa, in which the masseter muscle lies.

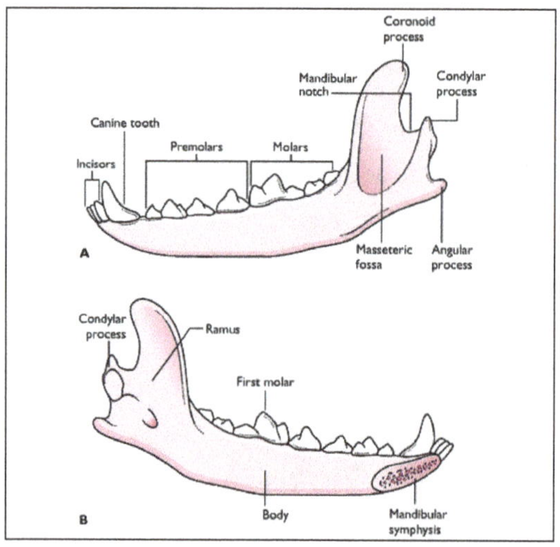

Lateral (A) and medial (B) views of the dog mandible.

Hyoid Apparatus

The hyoid apparatus lies in the intermandibular space and consists of a number of fine bones and cartilages joined together in an arrangement that resembles a trapeze. The hyoid apparatus is the means by which the larynx and tongue are suspended from the skull. The apparatus articulates with the temporal region of the skull in a cartilaginous joint.

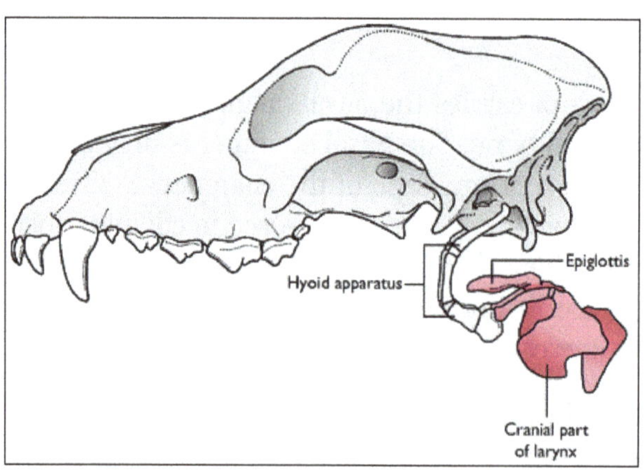

The trapeze-like hyoid apparatus suspended from the skull (dog).

Digestive System

In vertebrates, the digestive system is a hollow duct connecting the mouth, in the rostral part of the body, with the anus, in the caudal part of the body. Although there are major differences when comparing the digestive system between species, mostly due to the type of food they eat, we are going to describe the general organization of the digestive system of mammals having omnivorous diet.

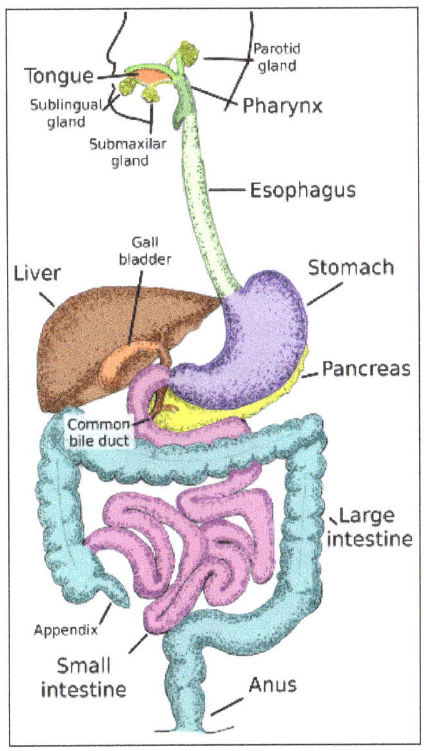

Drawing depicting the organization of the digestive system.

The digestive system is constituted of cephalic area, axial part and large glands (liver and pancreas). The cephalic area includes the oral cavity, salivary glands, teeth, as well as the tongue and the pharynx, which connects the digestive and respiratory systems, and bears the tonsils. The oral and nasal cavity is separated by the palate, excepting at the pharynx level. Both, the hard palate (anterior) and the soft palate (posterior) are covered by the same type of epithelium than the rest of oral cavity: stratified squamous epithelium. When comparing different animal species, the cephalic part shows the higher structural variability of the digestive system. The functions of the cephalic area are mechanical digestion, secretion of digestive enzymes like amylase, swallowing, as well as taste sensing.

The axial part of the digestive comprises the esophagus, stomach and intestine (small and large). Ingested food goes through the esophagus to the stomach. In the stomach, low pH and degradation enzymes produce partial digestion. After that, food moves to the small intestine for further break down. Water and nutrients are absorbed through the epithelium and into the blood and lymphatic vessels. Finally, leftover products are pushed into the large intestine and defecated through the anus.

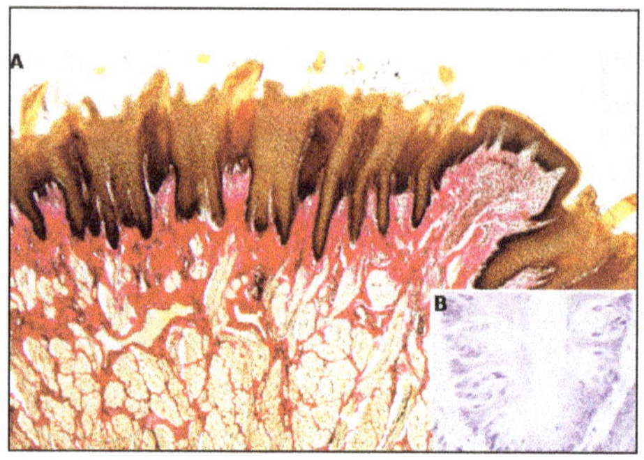

Tongue

The histological organization of esophagus, stomach and intestine is formed by four layers. The mucosa is the most internal layer, facing the lumen. It consists of an epithelium that covers the internal surface of the digestive tube, showing different morphological features related to protection, secretion and absorption functions, that it performs along the digestive tube. The epithelium rests on the basal lamina, and basal lamina rests on the lamina propria. Lamina propria is highly irrigated smooth connective tissue containing many cells of the immune system like macrophages, plasmatic cells, lymphocytes, and some others. Epithelium, basal lamina and lamina propria form the mucosa. In the lamina propria, there is a discontinuous layer of smooth muscle cells known as muscularis mucosa, located in the outermost part of the mucosa (far away from the epithelium). Muscularis mucosa makes possible the movement of the mucosa.

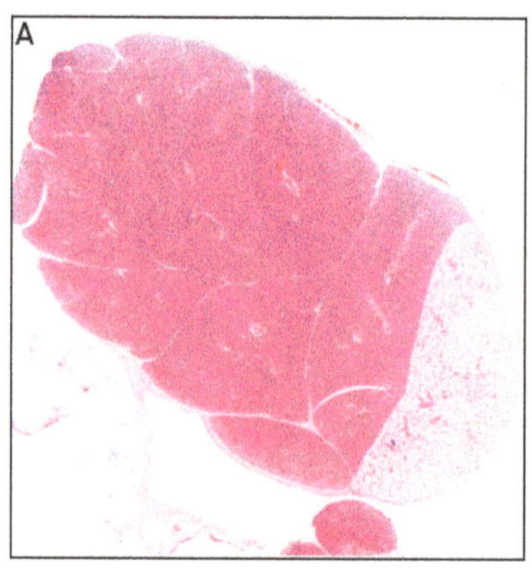

Salivary glands

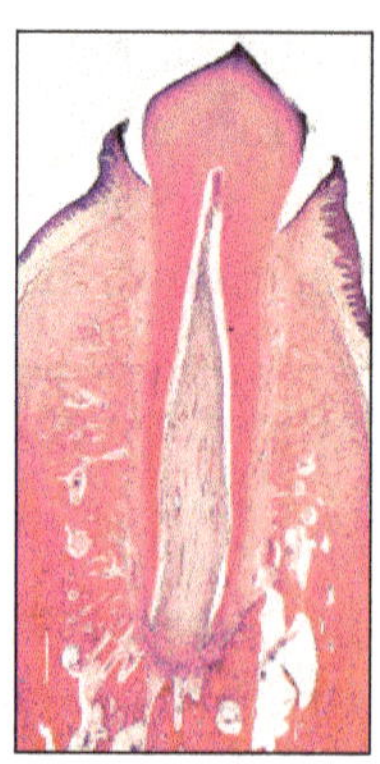

Tooth

The submucosa is the irregular dense connective tissue surrounding the mucosa. It contains many exocrine glands, also known as submucose exocrine glands. It is also highly irrigated by blood vessels and innervated by a dense network of nerves that form the Meissner's plexus, which controls the motility of the mucosa and the secretory activity of exocrine glands.

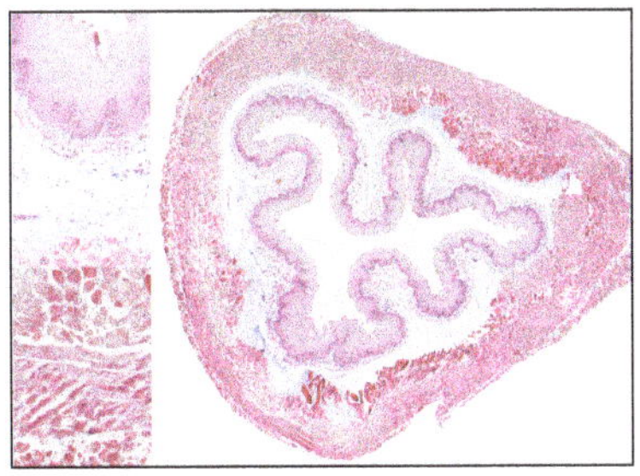

Esophagus

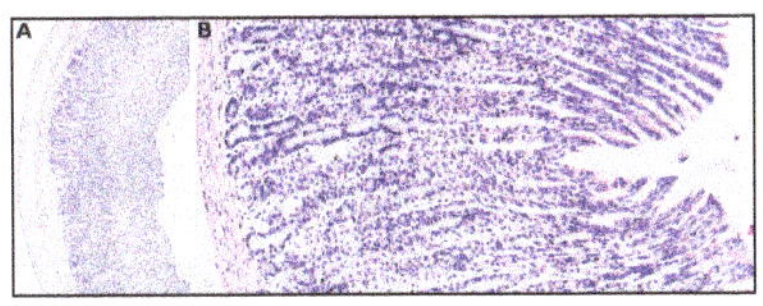

Stomach

The muscularis, or muscle layer, covers the submucosa. It is made up of smooth muscle cells, except the upper part of the esophagus that shows skeletal striated muscle fibers. Muscle layer is organized in two sublayers, an inner layer with circularly arranged cells, and an outer layer with longitudinally arranged cells. Between both layers, there is a neurovegetative neural plexus, known as Auerbach's plexus that drives the muscle layers contraction resulting in the peristaltic movements along the digestive tube.

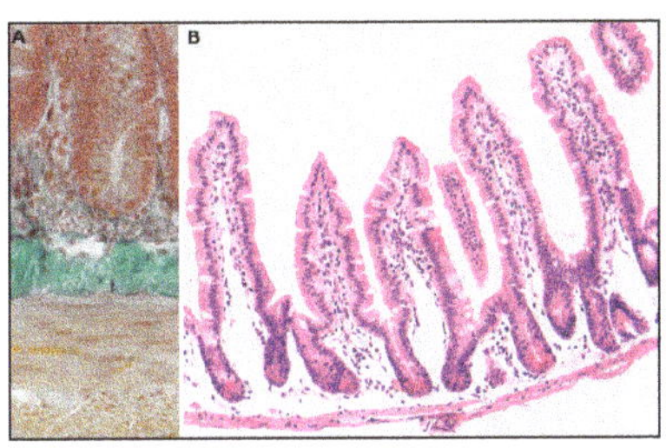

Small intestine

The serose, or adventitia, is the outer layer. It is made up of loose connective tissue with many adipose cells. This layer sets the border between the digestive tube and the coelomic cavities, and it is dorsally bound to the mesentery.

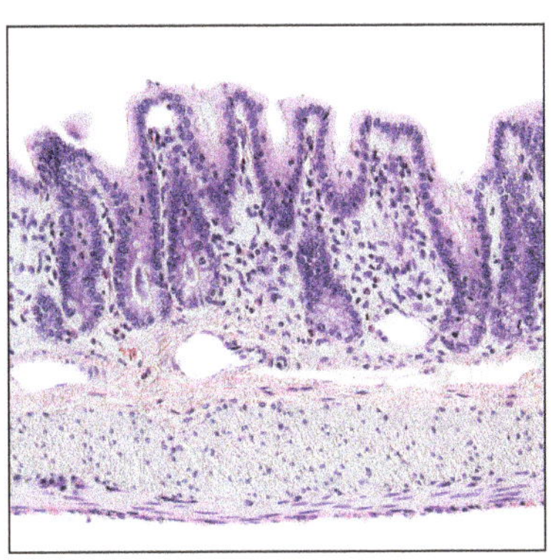

Large intestine

Two glands release their secretory products into de digestive tube: liver and pancreas. Both share the common bile duct as an excretory duct that ends up in the small intestine. Liver releases bile fluid containing biliary acids needed for digestion and absorption of fat. The exocrine part of the pancreas releases many enzymes for digestion. Both glands also have endocrine function by releasing hormones to the blood. Hepatocytes of the liver are able to synthesize and release both exocrine and endocrine compounds, whereas the endocrine and exocrine functions are performed by different parts in the pancreas. The Langerhans islets are the endocrine structures of the pancreas.

Digestive Systems in Animal Kingdom

Monogastric Digestive System

A monogastric digestive system has one simple stomach. The stomach secretes acid, resulting in a low pH of 1.5 to 2.5. The low pH destroys most bacteria and begins to break down the feed materials. Animals with this type of digestive system are better adapted to eat rations high in concentrates. Concentrates are highly digestible feedstuffs that are high in energy and low in fiber. Concentrates are typically 80 to 90 percent digestible. Common concentrates are cereal grains and oil meals. Cereal grains include corn, wheat, barley, and oats. Oil meals include soybean meal, linseed meal, and cottonseed meal. Examples of monogastric animals are hogs, cats, dogs, and humans.

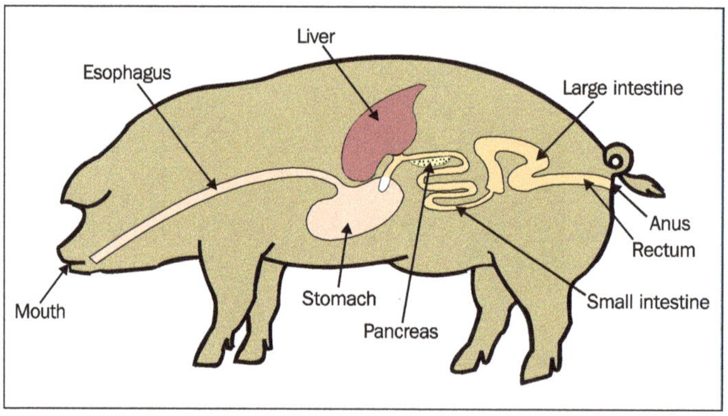

A basic diagram of the digestive system of a hog.

Avian Digestive System

The avian digestive system is found in poultry. This system differs greatly from any other type. Since poultry do not have teeth, there is no chewing. Poultry break their feed into pieces small enough to swallow by pecking with their beaks or scratching with their feet. Feed enters the mouth, travels to the esophagus, and empties directly into the crop. The crop is where the food is stored and soaked. Food then moves from the crop to the proventriculus. The proventriculus is the stomach in a bird, where gastric enzymes and hydrochloric acid are secreted. From the proventriculus, the food makes its way to the gizzard. The gizzard is a very muscular organ, which normally contains grit or stones that function like teeth to grind the food. The food then moves from the gizzard to the small intestine and then to the large intestine. The nondigestible food components then travel into the cloaca. Urine is also emptied into the cloaca. The material is then passed out of the body through the vent. Digestion in the avian system is very rapid.

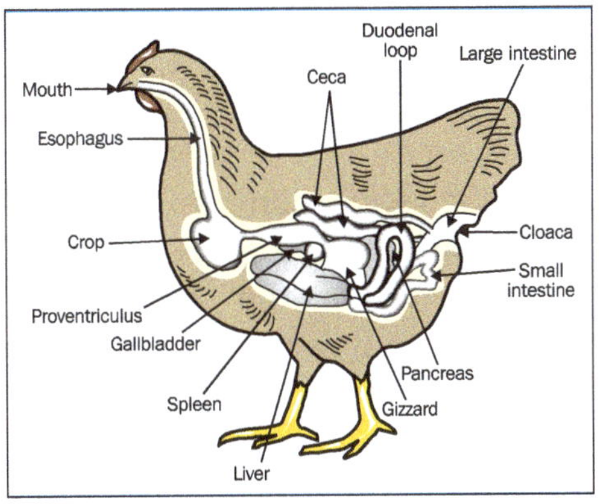

A basic diagram of the digestive system of a chicken.

Respiratory System

The organization and mechanisms allowing a chemical system to be materially and energetically connected with the environment, and equipped with the ability to self-construct, emerged first, and then appeared the complex chemical structures that provide the system with a temporal connection throughout successive generations. Thus, the origin of life was a process initiated within ecologically interconnected autonomous compartments that evolved into cells with hereditary and true Darwinian evolutionary capabilities.

Nevertheless there is a consensus that life started in an anaerobic environment in the so called "primordial broth", a mixture of organic molecules in the absence of oxygen. Molecular phylogenetic studies have revealed a tripartite division of the living world into two procaryotic groups, Bacteria and Archae, and one eukaryotic group, Eucarya. To know which group is the most "primitive" would help to delineate the characters of the last common ancestors to all living beings. According to several investigators and to the procaryotic dogma, the universal ancestor was probably a thermophile because primitive Earth was hotter than today. Nevertheless it is possible that the ancestor would have been a mesophile and, in this case, the root of the tree of life should be located in the eucaryal branch, with Archae and Bacteria sharing a common ancestor.

Almost four billion years ago, living beings that inhabited the earth were very primitive microorganisms, perhaps methanogenic bacteria, living in absolute anaerobiosis. These organisms still exist in our days and are included in the Archae domain, and for this reason are central to the paleoenvironment and paleobiology studies.

Anaerobic fermentation was a very inefficient metabolic process of extracting energy from organic molecules and the rise of an oxygenic environment was a momentous event in the diversification of life that dramatically shifted from inefficient to sophisticated oxygen dependent oxidizing ecosystems. Subsequently, oxygen became an indispensable factor for aerobic metabolism, especially in the higher life forms.

There are two widely accepted views of aerobic metabolism: First, that it was only possible after oxygen release by photosynthesis became abundant, and second, that it developed independently in diverse evolutionary lines. Analysis of the temporal distribution and geochemistry, suggest that the transition from reducing to sable oxygenic environment occurred later, between 2.3 and 1.8 billion years ago.

Molecular evidence shows that aerobic respiration evolved before oxygenic photosynthesis, or, in other words, cytochrome oxidase appeared before the water-splitting system. This hypothesis considers that denitrification (NO reductase) is the probable origin of aerobic respiration, that aerobic respiration arose only once the last universal ancestor was already present and that oxygenic photosynthesis developed in a single evolutionary line, the cyanobacteria, after the origin of aerobic respiration.

The rise of atmospheric oxygen caused by photosynthetic activity of evolving cyanobacteria must have created a remarkably strong selective pressure on organisms in both domains. Adaptations to use the new, chemically superior, electron receptor might have taken place, with similar molecular solutions creating the oxygen-reducing active sites. This would mean that the aerobic respiration has a single origin and may have evolved before the oxygen was released to the atmosphere by photosynthetic organisms and that the appearance of aerobic respiration was polyphyletic.

This new and more profitable method of extracting energy, aerobic respiration, should have led to a domain of aerobic organisms in the biological community, and probably even leading to the extinction of some anaerobic organisms.

The evolution to multicellular organisms determined the appearance of more sophisticated and specialized systems for the gas exchange, in order to develop an effective system of exchange. The existence of multicellular organisms, led to rapid and progressive morphofunctional differentiation of groups of cells that became the precursors of tissues, organs and systems currently present in more complex organisms. In most cases, this cellular specialization does not imply the loss of genetic material but only changes in the genes expression.

The respiratory systems from different groups of animals, although morphologically different, have in common the following characteristics: they have a large capillary network, the gas exchange surfaces are thin and moist; constant renewal of oxygen-rich fluid (air or water) order to provide oxygen and remove carbon dioxide; free movement of blood within the capillary network. According to Atwood, the assumptions stated above, are present either external respiration is carried out by 1) cutaneous diffusion (earthworm and some amphibians), 2) by thin tubes called tracheae (some insects); 3) by gills, the respiratory system of fish or 4) by diffusion through the lungs, respiratory organs present in amphibians, reptiles, birds and mammals.

Respiratory Organs in Vertebrates

The steps of the evolution of terrestrial vertebrates are: change from anaerobic to aerobic life, accretion of unicells into multicellular organs, formation of a closed circulatory system, evolution of metal-based carrier pigments that improved oxygen up-take, formation of invaginated respiratory organs, physical translocation from water to land, development of a double circulation and progression from ectothermic-heterothermy to endothermic-homeothermy. In vertebrates, the blood performs the task of transporting oxygen to the cells and carbon dioxide to the external environment. The development of the respiratory organs of vertebrates is closely related to the primitive pharynx, since the gills of aquatic vertebrates and the lungs of terrestrial vertebrates and aquatic mammals have pharyngeal embryology origin.

In all vertebrates, at a certain stage of their development, arise bilaterally in craneo-caudal direction from the inner side of the pharynx, a series of diverticula, which evaginate towards the outer surface, forming the pharyngeal pouches. The number of pharyngeal pouches is greater in lower vertebrates, reaching fourteen in cyclostomes and only four or five in birds and mammals. The pharyngeal pouches are separated by masses of mesenchyme that have the designation of pharyngeal arches, in which is located an arterial structure, called the aortic arch, which extends from the ventral aorta to the dorsal aorta.

During the ontogenesis of higher vertebrates, the pharyngeal pouches fail to open to the outside, contrarily to what happens in fish and, temporarily, in amphibians. Thus, in higher vertebrates, the pharyngeal pouches just remain during the embryonic period, where they undergo several changes, but very few or none of their initial characteristics are presented in adults. In amniotes, as in humans, only the first pair of pharyngeal pouches remains, giving origin bilaterally, the eustachian tube and middle ear.

Respiratory System in Fish

The fish gill adapted a structure for extraction of oxygen from water that is formed by a large number of filaments spaced out along the gill arches on either side of the pharynx. Each filament has a series of plates projecting at right angles from its upper and lower surfaces, the secondary lamellae, which are extremely numerous, are the site of gaseous exchange and form a fine sieve which ensures that all the water comes into close contact with the blood.

The gills are multifunctional organs that are responsible for the gas exchange (respiration) but also for the osmoregulation, acid-base regulation, and excretion of nitrogenous waste. The epithelial surface of a gill arch is structurally and functionally zoned. The filaments are covered by two distinct epithelial surfaces, the lamellar and filament epithelia, also termed the secondary and primary epithelia respectively. Gas exchange occurs through the secondary lamellae, and the non-respiratory functions of the gills take place in the primary epithelium.

The primary epithelium contains the chloride cells, which vary in morphology and number according to the milieu where the fish lives. The presence of an accessory cell beside the chloride cell is characteristic of seawater or seawateradapted fish. The secondary epithelium that covers the free part of the secondary lamellae has an exclusive relationship with the arterioarterial vasculature, i.e., the pillar cells. This epithelium consists of an outermost layer of pavement cells that exhibits structural characteristics suggestive of cell coat secretion and an innermost layer of less differentiated cells. In contrast to the primary epithelium, the secondary epithelium does not exhibit any obvious differences between freshwater and seawater fish.

The lamellar structure helps to increase the surface area but depends on the following complex anatomy to maintain the flat space necessary for circulation: separation between epithelial sheets (by pillar cells) and connection between the basal lamina of epithelial sheets by groups of strands (collagen columns).

To prevent ballooning and to ensure the sheet-flow dynamics of blood, the two layers of respiratory epithelium are connected by many strands of extracellular matrix (ECM) materials, which are called collagen columns. These columns, made of collagen fibers, are essential for reinforcing the lamellae structure and the internal force of blood pressure. Since collagen triggers the coagulation cascade when exposed to blood, the collagen columns are surrounded by the plasma membrane of pillar cells, which isolate them from circulation.

In the interface between pillar cells and collagen columns, exist adhesion junctions termed as "column junctions" and "autocelullar junctions", both of which are essential constituents of the gill lamellae. The "column junctions" is a cell-ECM adhesion and "autocelullar junctions" a membrane-membrane adhesion, both involved in maintaining structural integrity and hemodynamic of branchial lamellae.

The pillar cells have a spool-shape and possess a cylindrical cell body connecting two parallel sheets of respiratory epithelium. They also enfold 5 to 8 collagen columns and have numerous myofilaments, parallel to the collagen columns, which consist of actin and myosin, which form the contractile apparatus of the cell.

The pillar cells are a type of endothelial cells that delimits a network of vascular compartments within the lamellae of gill fish, but since they share characteristics with smooth muscle cells, we can say that these cells are specialized vascular cells with characteristics of both endothelial and smooth muscle cells.

The contractile apparatuses of the pillar cells possibly prevent collagen columns from being stretched and provide plasticity to the vascular network of the lamella against changes in blood pressure. Other possible function for the contractile structures of the pillar cells is that they can change the diameter of the vascular channels, and therefore contribute to the regulation of blood flow through the lamellae.

Besides the pillar cells, the gill epithelium of freshwater fishes have pavement cells (also termed as respiratory cells in older literature), mucus cells, neuro-epithelial cells and chloride cells. The neuroepithelial cells are isolated or clustered on the internal side of the primary lamellae facing the respiratory water flow. They are probably involved in local and central modulation of the branchial functions by interacting with the branchial nervous system and by paracrine secretion of substances such as serotonin. These cells share several morphofunctional features with the cells of the neuroepithelial bodies in the lungs of air-breathing vertebrates.

The chloride cells are described as large, granular, acidophilic and mitochondria-rich cells and exhibit an extensive tubular system emanating from the basolateral membrane, an array of sub-apical vesicles, large ovoid nucleus and abundance of Na^+, K^+ -ATPase enzyme . There is a marked difference between species in the structure of the apical membrane of chloride cells which precludes their absolute identification.

They are located in the primary epithelium in close proximity to the blood vessels and are sites of active chloride secretion and high ionic permeability, performing an integral role in acid-base regulation. As in other vertebrates, fish must maintain homeostasis of intra and extracellular pH and therefore use the parallel strategies of buffering and excretion to defend against pH changes. During alkalosis conditions, the area of exposed chloride cells is increased, which serves to enhance base equivalent excretion as the rate of Cl^-/ HCO_3^- exchange is increased. Conversely, during acidosis, the chloride cells surface area is diminished by an expansion of the adjacent pavement cells, and this response reduces the number of functional Cl-/HCO3- exchangers.

Under softwater or toxic conditions, chloride cells proliferate on both surfaces of the gill and might impair gas transfer owing to a thickening of the lamellar blood-to-water diffusion barrier. Water enters through the fish's mouth and out through the gill, slits in a direction that is opposite to the blood flow in the gill, providing a constant renewal of the oxygen supply in contact with the respiratory organ.

The exchange of oxygen and carbon dioxide takes place by diffusion from the surrounding water and the blood that flows within the capillary network of the gills, and because of this countercurrent flow fish can extract 80 to 90% of dissolved oxygen in water.

During the larval development of fish, the teleosts in particular, the skin is the cutaneous surface that ensures gas exchange, and only in the final stage of this period begins the hematosis through the gills, when the muscle-skeleton structure of the oral cavity becomes able to coordinate food intake with the flow of water through the branchial system.

Lungfishes

Changing conditions of life imposed new requirements on the morphology and physiology of the organisms. One of these changes is the evolutionary transition from aquatic to terrestrial life, leading to adaptations in locomotion, breathing, hearing, mechanism for food capture and other functions.

The first air-breathing vertebrates were fishes, and a Devonian air-breathing sarcopterygian (lobe-fin) occupies the basal position in the lineage extending from the Paleozoic fishes to the most derived tetrapods. The evolution of tetrapods from sarcopterygian fish is one of the major transformations in the history of life and involved numerous structural and functional innovations. The Styloichthys changae, one fossil of sarcopterygian fish, exhibits the character combination in a stem group close to the last common ancestor of tetrapods and lungfish.

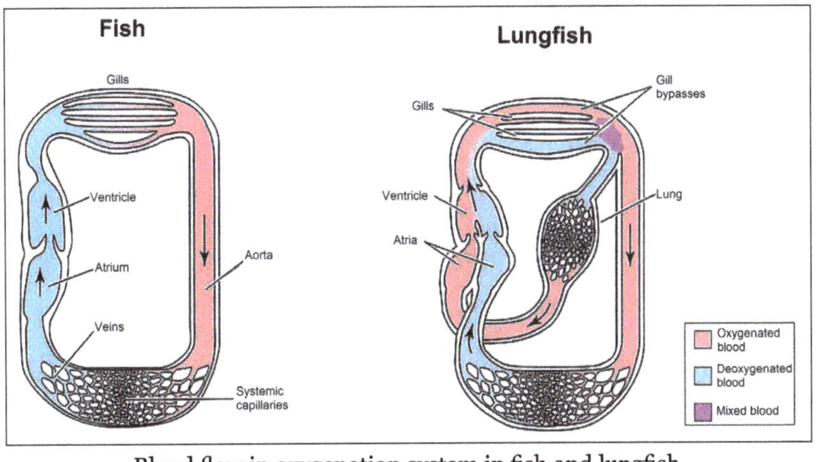

Blood flow in oxygenation system in fish and lungfish.

The recent discovery of a well-preserved species of fossil sarcopterygian fish form in the late Devonian of Artic Canada, that represents an intermediate between fish with fins and tetrapods with limbs, provides unique insights into how and in what order important tetrapod characters arose.

The morphological features and geological setting of this new animal are suggestive of life in shallow-water, marginal and subaerial habitats.

The relevance of the extant air-breathing fishes as models for events in the Paleozoic has been a recurring theme for more than one century. The lungfish is considered homologous to the lungs of all higher vertebrates and the precursor of the enteleost gas bladder.

In our days the lungfish are represented by three genera, the Australian lungfish, Neoceratodus forsteri, and the other two genera: the African (Protopterus) and South American (Lepidosiren) lungfish. The Australian Neoceratodus differ from the other lungfish because they breathe air for short periods and for this reason the lung is an accessory organ which is only used during periods of high activity in its natural habitat. They have efficient gills and possess only a single lung, unlike both Protopterus and Lepidosiren which have paired lungs and much reduced gills.

The lung of N. forsteri consists of a single elongated chamber compartmentalized by a thick cartilaginous structural framework. The epithelial lining of these supporting structures comprised abundant capillaries interspersed with cells resembling alveolar type II and type I cells. These epithelial cells which appear to be the only cell type lining the gas-diffusing surface, have long cytoplasmic plates bearing microvilli, which form part of the gas-exchange membrane. The cells contain large numbers of osmiophilic bodies resembling mammalian lamellar bodies, and it is possible that these lungfish cells may be the common ancestral cell for the alveolar type I and II found in the mammalian lung. They also have a surfactant-like material containing both SP-A and SP-B like proteins, suggesting that even in this primitive lung, these proteins are still involved in surfactant homeostasis.

To breathe air the Neoceratodus may raise to surface, exhale through the mouth, inhale and dive forward or rise to the surface, breathe and reverse back into the water. Protopterus and Lepidosiren are bimodal breathers, that use both gills and lungs for respiratory gas transfer, but they are obligate breathers because they die if denied access to air. The Protopterus occupy a variety of habitats both lentic (standing water) and lotic (running water) and possess the capacity to aestivate, reducing their metabolic rate, which allow them to survive to extreme heat or for long dry periods. In the Protopterus, the gills and skin uptake only 10% of the total O_2 uptake and these structures are much more effective in removing the CO_2.

The reedfish Calamoichthys calabaricus is one of the phylogenetically most primitive extant air-breathing fishes, and represents an animal successfully adapted not only for air breathing but also for making short-term use of terrestrial environments. In this primitive living actinopterygian fish the oxygen uptake is achieved by the gills, skin and a paired lungs and in the total volume oxygen uptake, the lungs account for 40%, the gills 28%, and the skin 32% .

In the Lepidosiren 99.15% of the total diffusing capacity lies in the lungs, 0.85% in the skin and only an insignificant 0.0013% in the gills, which shows that the gills don't have any importance as a gas exchange organ in this species? Oxygen uptake is accomplished by the lungs and dioxide carbon is eliminated by the skin.

The structure of the gills lamellae of Lepidosiren consists of a stratified epithelium that rests on the basal membrane and has at least three layers of cuboidal cells with large nuclei. Close to the epithelium there are numerous capillaries.

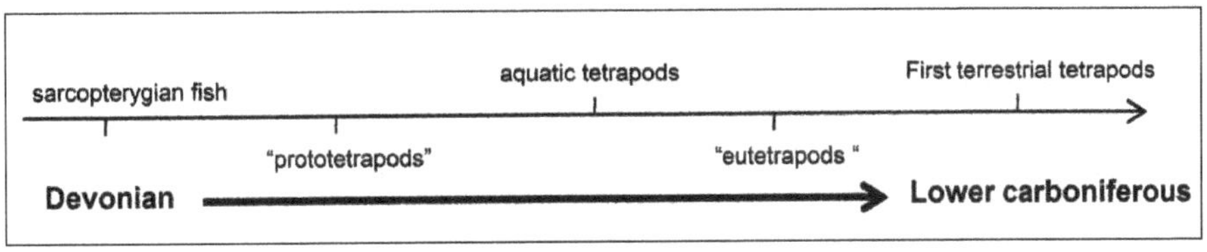

Tetrapods evolution.

Each lung of these lungfish has a main duct and numerous chambers of different sizes, which decrease in size as they progress caudally. The honeycomb-like edicular parenchyma is disposed in these chambers and most chambers contain a central lumen, which connects with the air duct. The duct, the chambers and edicular parenchyma consist of connective tissue septa held upright by smooth muscular/elastic trabeculae and are supplied and drained by branches of the pulmonary artery and vein. Most interedicular septa have a double capillary net. The air-blood barrier consists of three layers: a simple squamous epithelium made up of a single type of cell, the endothelial cells of the blood capillaries and the combined basal lamina of the epithelial and endothelial cells.

The skin of these fish has two layers, the epidermis and the dermis. The epidermis consists of a stratified epithelium with six to ten layers of diverse cell types. Most prominent are superficially located cuboidal cells with a large central nucleus and the mucous cells that are dispersed among the other cell layers. The dermis is a dense connective tissue, with blood vessels and small ossified scales. Numerous blood capillaries and melanophores lie beneath the basal membrane and between the subjacent layers.

In the Protopterus and Lepidosiren the ventilation of the gills occurs through the action of a positive pressure buccal pump anterior to the gills and an opercular suction pump posterior to the gills. These pumps generate a nearly continuous water pressure gradient favouring a water flow in the mouth, through the gills and out the opercular opening.

The ventilation of the lungs in Protopterus is achieved by the same musculoskeletal elements involved in aquatic ventilation, the buccal force pump mechanisms. Early in their history, fish developed supplementary air breathing organs in two taxonomic lines – Actinopterygians and Sarcopterygians. The onset of aerial respiration in primitive fish was an important milestone in the evolution of terrestrial vertebrates. The fish-tetrapod transition was one of the greatest events in the vertebrate evolution. Tetrapods first appeared in the late Devonian about 360 million years ago, but appear to have been primarily aquatic animals. For some investigators the freshwater origin of tetrapods remains the most likely scenario, but several recent findings raise the possibility that the tetrapod land invasion could come from a marine habitat.

The evolution of tetrapods occured under environmental influences and presumption that hypoxia habitat conditions were similar to those commonly encountered in tropical lowland habitats during dry seasons.

The sequence of evolution begun with sarcopterygian fish, followed by the appearance of a "prototetrapods" (e.g. Elginerpeton), the emergence of aquatic tetrapods (e.g. Acanthostega), the appearance of "eutetrapods (e.g. Tulerpeton) and the first truly terrestrial tetrapods (e.g. Pederpes) in the lower Carboniferous. Several morphological changes were observed during the evolution process, developing specialized features that allowed land locomotion and air breathing.

The sudden change from gill respiration to lung breathing would pose considerable physiological problems. One of the consequences of gill loss, would be the concentration of respiratory CO_2 within the body, which required buffering by bicarbonate ion and affected processes such as acid-base balance, O_2 binding by haemoglobin, ventilation rate, respiratory control and also affected nitrogen excretion, ion regulation and water balance, vital processes that would need to be assumed by others organs. The advantages of tetrapod gill loss included head mobility, development of hearing and the origin of different ventilatory and feeding mechanisms.

In the primitive pure buccal pumping, found in most airbreathing fishes, including lungfishes, the axial musculature does not contribute to expiration or inspiration. In fact, the buccal pump breathing has been proposed to constrain the evolution of tongue morphology and head shape. The aspiration breathing was present in some early tetrapods, but it only arised when the early amniotes appeared. Aspiration breathing evolved in two steps: first, from pure buccal pump breathing to the use of axial muscles for expiration and buccal pump for inspiration; second, to pure aspiration-breathing, in which axial muscles are used for both expiration and inspiration.

The musculoskeletal units responsible for breathing also serve other functions such as feeding or locomotion, and the conflicting mechanical requirements of multiple functions possibly constrain the performance and evolution of one or both functions. The evolution of aspiration breathing may have allowed the musculoskeletal systems of the head and tongue of amniotes to diversify, but the ribs and intercostal musculature became constrained by their dual function in aspiration breathing and high speed locomotion.

The loss of gills is also in connexion to the cutaneous respiration as a site gaseous exchange which can function in water and land. So, the possible evolution of the respiratory mechanisms maybe begun with an ancestral fish adapted for oxygen uptake and CO_2 elimination in aquatic medium, but under conditions of low oxygen, developed adaptations that allowed the fish to come to the surface to obtain extra oxygen from the air, but the gills still functioned for CO_2 elimination.

At this stage of evolution, the skin would function mainly for CO_2 elimination and, as the lung became more efficient and more involved, not only in uptake of oxygen but also in elimination of carbon dioxide, the skin became less important and probably covered with hardened scales to reduce water loss and the animal could now remain away from water for longer periods. It seems possible to accept that the cutaneous respiration was important for the earliest land vertebrates.

Respiratory Organs in Amphibians

Based on paleontological criteria, the first amphibians have arisen by evolution of fish Crossopterígeos ripidistios, extinct in the late Devonian period. Modern amphibians occupy a central position in understanding the fundamental changes that have occurred in the evolution of air breathing. Dual subsistence in water and land has required development of certain respiratory adaptations.

The transition from aquatic to land environment exposed the gas exchange organ to a much richer oxygen ambience, which allowed a drastic reduction in the ventilation requirements, but at the same time created problems for the disposal of carbon dioxide, because at 20 °C the water solubility of this gas is 28 times greater than that of oxygen.

To prevent a severe respiratory acidosis, the Terran animal began to use the skin as an important respiratory organ, designed especially for the removal of carbon dioxide, which required a substantially reduction of the barrier represented by the scales that covered the surface of their aquatic ancestors. At the same time there must have occurred an increased bicarbonate concentration in plasma, in order to compensate the increase of carbon dioxide.

These animals are mainly characterized for presenting an aquatic larval form, the tadpole stage, where hematosis takes place through the gills. Next they suffered a metamorphosis that allowed them to reach adulthood in terrestrial habitat and in which the breathing air was carried out by the lungs, skin and mouth. The amount of cutaneous and buccal gas exchange and its percentage in the total gas exchange, varied from species to species and also during seasons.

Amphibians have the simplest lungs, rudimentary lungs that are adequate for ectothermic and low aerobic metabolism animals. The paired lungs of recent amphibians are unicameral lying in the dorsal pleuroperitoneal cavity. In the various amphibian species the lungs differ greatly in size, their topographic extension and the dimension of exchange surface by the development of interconnected folds with highly varying number of subdivisions and height of their folds. The highly varying extent in lung exchange is due to differences in the amount of gas exchange performed by via lungs in concert with cutaneous and buccal cavity exchange.

Moreover, the absence of an individualized chest well, with no ribs or diaphragm, the amphibian's pulmonary ventilation is mainly accomplished at the expense of swallowing air, carried out by rising of the oral cavity floor.

The remarkable heterogeneity of the morphology of the amphibian gas exchangers matches that of the diversity of the environments in which the animals live, the lifestyle they pursue, and their pattern of interrupted development. The skin is the main pathway for gas transfer in aquatic species while in terrestrial ones; it has been relegated or rendered redundant.

In the salamanders (Plethodontidae), some of which live in cold well-aerated waters, gas exchange occurs across the skin and buccal cavity. Skin breathing is important in all extant amphibians but is the only means of gas exchange in those salamanders (terrestrial and aquatic) which possess neither lungs nor gills. Gas exchange takes place in the dense subepithelial capillary network, the inflow to which is in part from the arterial system and in part from a branch of the pulmonary arch carrying venous blood. The oxygenated cutaneous blood flows into the venous system. This is in contrast to the arrangement of pulmonary outflow in tetrapods and lungfish which allows (complete or partial) separation of oxygenated from venous blood.

The caecilians (Apoda) possess long, tubular lungs, but in some species the left lung is remarkably reduced or totally missing. The lungs of caecilians are internally subdivided, forming air cells that are supported by diametrically placed trabeculae.

In the newts (Urodela), animals that are mostly aquatic, the lungs are poorly vascularised with the internal surface being smooth. Lungs of most amphibians such as Amphiuma tridactum and the cane toad, Bufo marinus, have an abundance of smooth muscle tissue, a feature that may explain the high compliance of the lungs. In Amphiuma, during expiration, the lung virtually collapses, producing an almost 100% turn-over of inspired air. Amphiuma is aquatic but has very well developed lungs.

The lungs of terrestrial species are highly elaborate presenting a series of stratified septa that divided the lung into small air cells and the lungs of Anura and Apoda are more complex than those of Urodela.

Respiratory Organs in Reptiles

It is assumed that reptiles made their appearance on Earth about 310 million years ago, and their adaptation was so perfect that they dominated the planet for over a hundred million years. The innumerous fossils that have been discovered allow us to group them in a numerous orders capable for living in different habitats, such as land, air or aquatic environment.

Reptilians are the first vertebrates adequately adapted for terrestrial habitation and utilization of lungs as a sole pathway for acquisition of oxygen. The skin that was no longer necessary for gas exchange became an armor to protect against dehydration, being waterproof, dry, covered with keratinized epidermal scales or developing dermal bone plates.

Compared with their gigantic prehistoric ancestors, current reptiles are small and insignificant and can be grouped in four orders: chelonians, such as turtles and tortoises; rincocéfalos, like Sphenodon of New Zealand; crocodiles (crocodiles and caimans) and squamata order (lizards and snakes).

The reptilian display great pulmonary structural heterogeneity and there is no single model of reptilian lung. Based on complexity of internal organization, different classification suggested that the turtles, monitor lizard, crocodiles and snakes have a profusely subdivided (multicameral) lung, the chameleons and iguanids have a simpler (paucicameral) lung and the teju lizard (Tupinambis nigropunctatus) have a saccular, smooth-walled, transparent (unicameral) lung.

Division of the lumen of the lung into a number of chambers, by septation, enlarges the exchange area, fact that is observed in turtles, lizards and crocodiles. The lungs are localized in the pleuroperitoneal cavity and there is no diaphragm separating the thoracic from the abdominal cavity. Presence of ribs and intercostal muscles in reptiles, allow the development of more effective pulmonary ventilation than that of the amphibians which do not have these anatomical structures.

Generally, the pattern of organization of the respiratory system of reptiles is identical to mammals, with the lungs coated externally by a serosa. The conducting portions are supported by complete cartilaginous rings, which continue through the extra and intrapulmonary bronchi. The branching of the bronchial intrapulmonary tree in reptiles is similar to mammals'; however they have specific designations, which appear sequentially bronchus, tubular chambers, niches and aedicules.

The intrapulmonary bronchi of the reptiles that give immediate access to respiratory areas correspond to the mammalian respiratory bronchioles, the tubular chambers, according to their position and morphofunctional structure, are equivalent to the alveolar channels in mammals, and the niche are similar to alveolar sacs. By its position in the respiratory system and anatomical constitution the aedicules are equivalent to the alveoli of mammals, however they have an oblong structure compared with the spherical form of mammal's alveoli.

The intrapulmonary bronchi of turtles that live essentially in aquatic environment have a reinforcement that extends to or near the respiratory areas, characteristic that is similar to the aquatic mammals that have the ability to dive to great depths, such as seals, dolphins and whales. This reinforcement, along with the presence of a smooth muscle, appears to be adaptations that allow these animals to support the high pressures to which they are subjected during the immersion to great depths.

The epithelium of the trachea and bronchi is pseudostratified columnar ciliated, with non-ciliated secretory cells and basal cells, all in direct contact with the basal membrane. Isolated or groups of neuroendocrine cells were also identified within the conducting portion of the lung of turtles and crocodiles.

The epithelial cells lining the respiratory surface of reptilian lungs are differentiated into type I and type II cells and it is possible to observe multilamellar bodies similar to those present in mammals. These suggest that also in reptiles occurs the synthesis of surfactant lipoprotein material responsible for the stability of their respiratory unit, the aedicula.

The role of surfactant in reptiles, which are not highly susceptible to collapse from surface tension forces, is obscure, and may have other important functions such as prevention of transendothelial transudation of blood plasma across the blood-gas barrier, immune suppression and attraction of macrophages. Reptilian lungs have preponderance of smooth muscle tissue and this tissue has been associated with intrapulmonary connective movement of air.

Respiratory System in Birds

Birds' respiratory system, the lung – air sac system, is the most complex and efficient gas exchanger that has evolved in air-breathing vertebrates. The compact and virtually constant-volume avian lung has been totally uncoupled from the compliant, avascular air sacs.

The main properties that impart high respiratory efficiency on the lung–air sac system of birds are a cross-current design and inbuilt multicapillary serial arterialisation system; auxiliary counter-current system; large tidal volume; large cardiac output; continuous and unidirectional parabronchial ventilation; short pulmonary circulatory time; superior morphometric parameters ; a particularly large respiratory surface area and a remarkably thin blood-gas (tissue) barrier.

Their respiratory system allows them to breathe at altitudes that can reach nine thousand meters without acclimatization, fact that is impossible for humans and other mammals, in which the barometric pressure of high altitudes can at least induce a comatose state.

To maintain physiological function at high altitude, under reduced environmental oxygen availability, the capacity to transport O_2 must increase. The exposure to hypoxia causes an immediate increase in breathing due to stimulation of arterial chemoreceptors and changes in metabolic state. The ability to adjust peripheral heat dissipation to facilitate the depression of body temperature during hypoxia, which reduced the metabolic demand, allows birds to fly high and for long periods, and is a result of an evolutionary adaptation. The bar-headed goose, a typical high altitude bird, depresses metabolism less than low-altitude birds during hypoxia and breathes substantially more than birds that fly at low altitudes. The bar-headed goose has hemoglobin with higher O_2 affinity and may be capable of generating higher inspiratory airflows.

The respiratory system of birds is separated into lung (the gas exchanging part) and a series of airs sacs (non-respiratory) with anastomosing air capillaries and pneumatized bones that allow unidirectional flow of air compared to the blind sac and tidal flow in mammalian lungs.

Lack of diaphragm displaced the lungs to the coelomic cavity where they are closely attached to the ribs. Intercalated between the sacs, the lungs are largely continuously ventilated back-to-front by a concerted action of the cranial and caudal groups of air sacs.

There are fundamental differences in the breathing mechanics of different birds, driven in part by the morphological differences of the rib cage and sternum associated with skeletal adaptations to locomotion. The uncinate processes are bony projections that extend from the vertebral ribs, providing attachment sites for respiratory muscles.

The elongation of ribs, rib cage and sternum associated with diving species, as well as longer uncinates, maybe important upon resurfacing when inspiration occurs against the pressure of water against the body. The reduction in the sternum and the shortest uncinate length found in the walking species, suggests that they may play a reduced role during breathing in these species.

The circular lumen of the trachea has a cartilaginous or partially ossified support ring, whose number varies according to species and is lined by a cilindric pseudociliated epithelium with goblet cells. The trachea bifurcates into two primary bronchi, with an epithelial lining similar to the trachea but with incomplete cartilaginous rings, which disappear or are reduced when they reach the bronchial lung parenchyma.

The pair of lungs of the birds are relatively small, noncompliant, localized in the dorsal thorax region and with little moving during breathing, as air is driven unidirectionally though the lung, via the system of air sacs.

The connection between the primary bronchus and the secondary and tertiary bronchi is labyrinthic, markedly opposed to the monopodic branch of mammal. The primary bronchus gives rise to four craneo-medium secondary bronchus and to seven caudal-dorsal secondary bronchus. In the secondary bronchus the mucosa is lined by a simple cuboidal or columnar epithelium, without goblet cells. Tertiary bronchus or parabronchus are arranged in a series of parallel lines, whose ends are open to the secondary bronchus. All the way through the parabronchus have recurrent anastomoses between them. The number of parabronchus varies from species to species, but is higher in the birds that fly better, has been estimated in Gallus domesticus between 300 and 500 parabronchus. Parabronchus have an average diameter of 500 μm, and are lined by a simple squamous epithelium, just like the mammals' alveolar channels. Along the inner surface of parabronchus, small vesicular structures with hexagonal shape emerge, with 100 to 200 μm in diameter. These structures called atria are separated from each other by septa mainly consisting of smooth muscle cells located in the freeboard, and collagen and elastic fibers, located at the base. The atria epithelium has two types of cells, one of which are the granular cells that are confined to the atria, have a cytoplasm that contains multilamellar bodies and are considered analogous to the cells of type II pneumocytes of the mammal's lungs . The other are the squamous cells, which line the inner surface of the atria andare based on a continuous basal membrane, forming the simple squamous epithelium. From the deepest area of each atrium arise 2 to 4 infundibula that continue with the air capillaries with 3 to 10 μm diameter. They are lined by squamous cells, that are similar to the cells of the atria, but they are not based on the basal membrane. The infundibula and air capillaries of adjacent atria form anastomosis to one another.

The blood capillaries are surrounded by extremely small air capillaries and other capillaries, which give an appearance of a dense network. The blood capillaries are embedded in a rigid structure with numerous cross-braces that provide mechanical support of the small vessels at numerous points. This feature contributes to mechanical strength of blood capillaries and allows them to have a remarkably thin blood-gas barrier (BGB) that is uniformly arranged all around the circumference of the blood capillary.

The diameters of the air capillaries are comparable to those of blood capillaries and as a consequence of the very small diameter, the surface tension of these air capillaries is so high, despite their very well-differentiated surfactant, that they can only remain patent as rigid structures in a volumeconstant lung. The surfactant of these rigid air capillaries lowers the high air capillary surface tension to such an extent that the remaining surface tension cannot suck fluid from the blood into the air capillary, thus preventing edema and maintaining gas exchange.

Together with the extensive network of blood capillaries, the air capillaries form the gas exchange surface of the bird's lungs. Unlike those observed in lung alveoli of mammals, the air capillaries are not terminal fund sacs formations, and therefore allow an unidirectional air flow through the lungs of birds.

The lung air sacs are pair formations, and their total number for the two lungs varies between 6 and 14 depending on the species, are generally referred to as cranial group and caudal group, and all cranial bags communicate with all secondary bronchi, fact that does not occur with caudal bags. The oxygen concentration is higher inside the caudal bags whereas the concentration of carbon dioxide reaches higher values inside the cranial bags. This qualitative difference is explained by the particular pattern of unidirectional airflow that occurs in the lungs of birds.

The inhaled air moves into the respiratory system, whenever the chest cavity expands by the action of inspiratory muscles, and during expiration the air is expelled by action of the expiratory muscles. Although the birds do not have diaphragm, the entry and exit of air in to the respiratory system is a process similar to the observed in mammals.

During inhalation, air flows through the mesobronchus in to the posterior air sacs, and at the same time, the air enters the anterior air sacs via the dorsal secondary bronchus and parabronchus. During exhalation, the air leaves the posterior air sacs and passes through the parabronchus, and to a lesser extent, through the mesobronchus, to the trachea. At the same time in the anterior air sacs, the air moves through the secondary ventral bronchus towards the trachea. There is thus, during the two phases of the respiratory cycle, a continuous unidirectional flow through the parabronchus.

The flow inside the air capillaries and inside the blood capillaries occurs in opposite directions, i.e., in the blood capillaries the flow goes from the most peripheral areas of the parabronchus into its axis, while in the air capillaries move away from the axis of parabronchus. It is established a counter-current system of gas exchange between air capillaries and blood capillaries.

Briefly we can say that the anatomical and physiological features that have just been summarily exposed, such as a continuous and unidirectional ventilation of parabronchus, lower capacity of the air sacs, large surface for gas exchange and a very thin air-blood barrier, explain the unique characteristics of the respiratory system of birds, allowing them to extract oxygen from the highly rarefied air atmosphere.

Respiratory System in Mammals

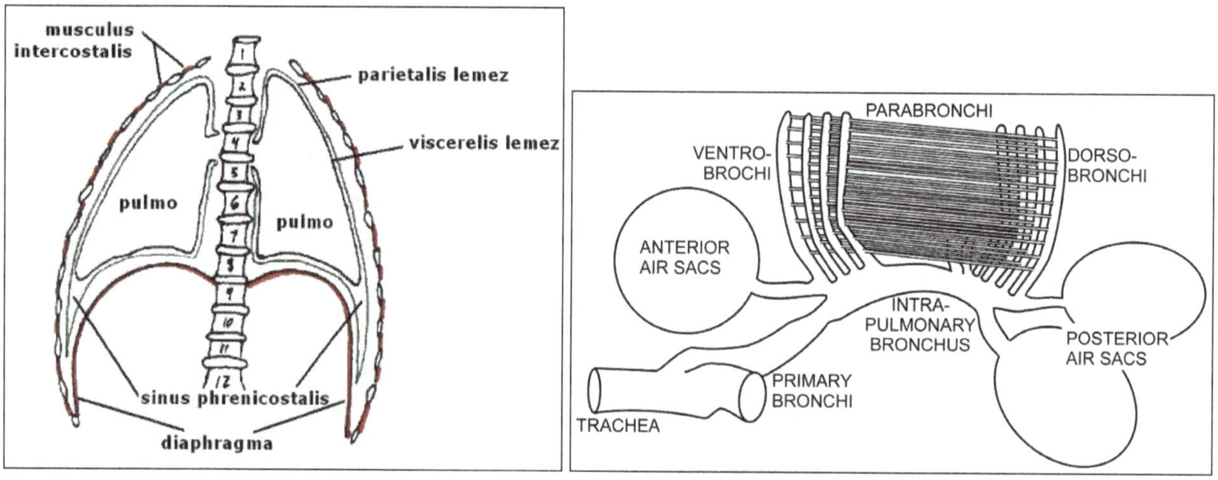

Comparation between the birds and mammals respiratory systems.

Some 300 million years ago, the ancestors of modern reptiles finally emerged completely from water and made a commitment to air breathing. From them, developed the two great classes of vertebrates with high maximal oxygen consumption: the mammals and the birds. A remarkable feature of these two groups is that although the cardiovascular, renal, gastrointestinal, endocrine and nervous systemsshow many similarities, the lungs are radically different. The distinct morphology of avian and mammalian lungs reflects not only an increased demand for gas exchange, but is historically correlated with the divergent modes of locomotion that facilitate higher rates of ventilations.

It is thought that mammals made their appearance on Earth during the Jurassic Period, the age of reptiles, when the process of divergence of the continents begun. Mammals evolved homoiothermy independently from birds, but in a very similar way. For the mandatory increased metabolism, they required a correspondingly increased gas exchange surface, which became available by the development of the broncho-alveolar lung.

The nearest ancestors of mammals appear to have been same group of reptiles and the lung of the mammals derived from a multicameral reptilian lung with three rows of lung chambers. The branched conducting bronchial system originated by stepwise further subdivision of these lung chambers, terminating in the branched respiratory bronchioli and ductus, covered with alveoli.

Of all tetrapods' breathing systems, the mammals' respiratory system has been the most extensively studied, often with the aim to acquire knowledge with medical relevance. In mammals there is no dissociation between locomotion and respiratory movements and both are closely coupled especially during exercise.

The strong musculature of the diaphragm does not only act as a forceful inspiratory muscle together with the intercostals musculature, but is also responsible for maintaining a pressure gradient between the pleural and the peritoneal activity during strong exercise. During respiration at rest, expiration is performed by elastic retractile forces of the extended rib cage and by the retraction forces of the lung itself out of the surface tension of the alveoli together with their extended elastic fibre systems. During exercise, expiratory movement of the intercostals musculature is strongly supported by the muscles of the abdominal wall, which is also the case for all sound productions, speech and singing.

In mammals the lungs do not empty completely during the expiration, and the result is that convective flow alone cannot take the inspired gas to the periphery of the lung where some of the gas-exchanging alveoli are located. Instead the last part of the distance is accomplished by a relatively large peripheral airways to allow mixing of the inspired air with that already in the lung, and the resulting large alveoli cause additional problems.

In the mammalian lung, the airway and vascular systems form a complex multigenerational dichotomous branching tree-like arrangement. Transported by bulk-flow (convention) in the initial (large) parts of the bronchial system and mainly by diffusion in the terminal (fine) sections of the airway system, the inspired air ultimately reaches the alveoli where it is exposed to capillary blood across a thin, extensive tissue barrier.

The alveolar surface is mainly lined by type I and type II cells. Type II cells secrete surfactant. In mammals the capillaries are located in the alveolar walls which are widely separated from each other. Thus the BGB has to withstand the full transmural pressure. The capillary is typically polarized with one side having very thin BGB whereas on the other side the barrier is thicker and contains strands of type collagen which provides support for the alveolar wall and maintains the integrity of the alveoli . In contrast to a uniform thin BGB in the birds, in mammals half of the surface area of the capillaries provides inefficient gas exchange due to its increased thickness.

Excretory System

The excretory system consists of several organs and structures that may have diferent features depending on the animal species. Its main function is to eliminate by filtration waste products resulting from cellular metabolism or incorporated from the environment, for example, during feeding, salt excess in seabirds, etcetera. The main excretory system in animals is the urinary system.

The urinary system is composed of the kidneys, in which urine is produced, ureters, urinary bladder, and the urethra. An initial filtration of the blood is produced in the kidneys. Some of the molecules from this filtration are later reabsorted and returned to the blood vessels. After this process, the resulting liquid, called urine, is collected in tubes of increasing diameter: the collecting ducts, the calyces, the renal pelvis and the ureters. The ureters arise from the kidneys and transport urine to the urinary bladder, which is a temporary storage compartment. Finally, stemming from the urinary bladder, the urethra is the duct through which the urine is evacuated outside the body.

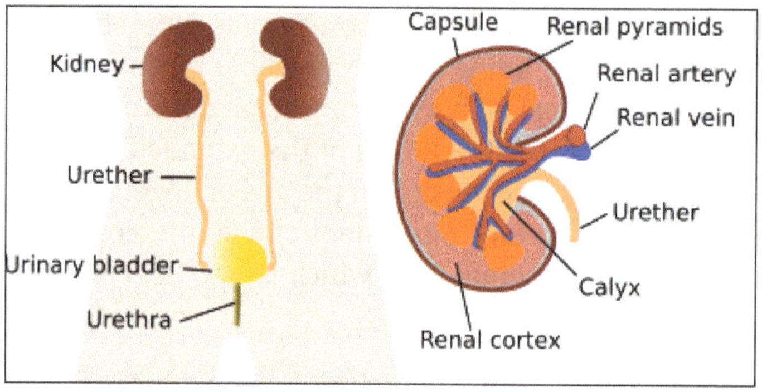

Urinay system.

The kidneys are covered by a connective fibrous capsule under which the cortex is located. The medulla is the inner part of the kidney. Cone-shaped structures called renal pyramids are observed in the medulla.

The structural and physiological unit of the kidney is the nephron. In fact, the kidney is made up of many repetitions of this structure. The nephron consists of the renal corpuscle, where the first filtration of the blood takes place. This filtrated solution is channeled to the renal tubule. Several segments can be distinguished in the renal tubule: a proximal tubule (divided into convoluted and straight portions), an intermediate tubule (divided into descending and ascending regions), a distal tubule (divided into convoluted and straight portions) and a connecting tubule. The straight portion of the proximal tubule, the intermediate tubule and the straight portion of the distal tubule form the so called loop of Henle.

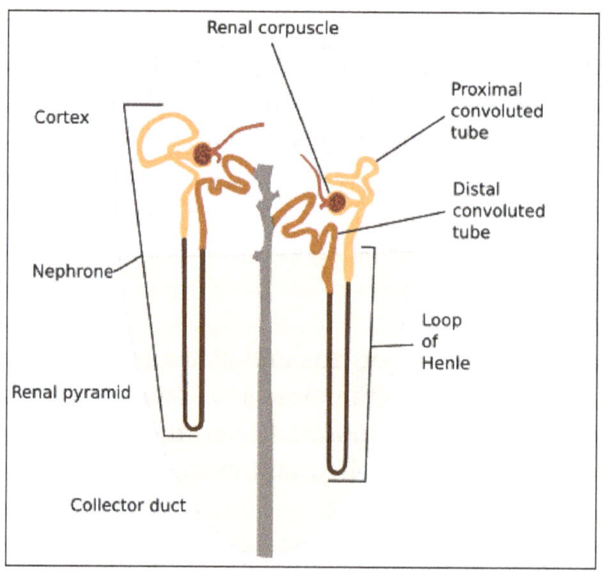

Nephrone.

The distribution of nephrons and their tubules follow a highly organized model that can be seen with a magnifying glass or macroscopically. As a result, the glomeruli and the convoluted portions of the proximal and distal tubules form the renal cortex. At the same time, the Henle's loop and the collecting ducts confer a typical organization to the medulla by forming the renal pyramids.

The renal vascular system is quite complex. The renal artery enters through the renal hilumand branches out profusely in afferent arterioles before forming the glomerular capillary tuft. The efferent blood vessel is not a vein but an efferent arteriole, which leaves the glomerulus by a point near to the afferent arteriole. These arterioles form straight vessels that irrigate the medulla and then continue with the venous rectus vessels. All of them gather in increasingly larger veins to exit the kidney through the renal hilum in the form of a renal vein.

The urinary system, besides the nephrone, is formed by all the hollow ducts and organs that transport or store urine, including the renal pelvis, the ureter, the urinary bladder and the urethra. All of them, with the exception of the medial and distal portion of the urethra, are compose of a mucosa formed by a transitional epithelium (urothelium) and by a subepithelial connective tissue. The transitional epithelium is characterized by its adaptation to changes in volume that occur mostly in the urinary bladder. A smooth muscle layer is found below the mucosa, and finally a wrapping layer of connective tissue known as adventitia.

The urinary bladder is a container for urine, in which the ureters from each of the kidneys converge, and from which the urethra emerges. The wall of the urinary bladder has the capacity to form folds when it is empty, and to stretch when it is filled.

The urinary ducts show sexual differences, especially concerning the urethra. While in the woman the urethra is a tube for expeling only the urine out of the body, in the man it also has to expel the sperm. That is why it is also known as the uroseminal duct. This functional difference implies a difference in length: 3-4 cm for women and 20-25 cm for men. In addition, there are differences in the type of epithelium that covers the interior of the urethra, and in the associated glands.

Circulatory System

In all animals, except a few simple types, the circulatory system is used to transport nutrients and gases through the body. Simple diffusion allows some water, nutrient, waste, and gas exchange into primitive animals that are only a few cell layers thick; however, bulk flow is the only method by which the entire body of larger more complex organisms is accessed. The heart, veins, capillaries, arteries, lymph vessels, and lymph glands comprise the circulatory system of domestic animals. Together, they work to supply the body tissues with nourishment and to collect the waste materials of the body. Specific Functions of the Circulatory System:

- Distribution of nutrients.
- Transportation and exchange of oxygen and carbon dioxide.
- Removal of waste materials.
- Distribution of endocrine secretions.
- Prevention of excessive bleeding.
- Prevention of infection.
- Regulation of body temperature.

Anatomy and Physiology of the Heart

The heart, which is located near the center of the thoracic cavity between the lungs, is responsible for pumping blood to all body parts. The heart is a funnelshaped, hollow, muscular organ that is housed in a sac called the pericardial sac. The pericardial sac consists of two layers giving support to the heart, and it contains a minimum amount of fluid for lubrication. Further support is given to the heart at the base, or broad end, by large arteries and veins. The apex, or pointed end of the heart, is directed toward the abdomen.

The wall of the heart consists of three layers: epicardium, endocardium, and myocardium. The outer layer, or epicardium, is actually the inside layer of the epicardial sac. The endocardium, or inner layer, consists of endothelial cells.

It lines the heart, covers the heart valves, and serves as the lining of the blood vessels. Cardiac muscle, which is involuntary, striated muscle, is the composition of the myocardium, or middle layer, of the walls of the heart. The muscle cells of the heart are striated, but the nucleus is more centrally located than it is in voluntary, striated muscle cells. The muscle fibers of the heart tend to intertwine.

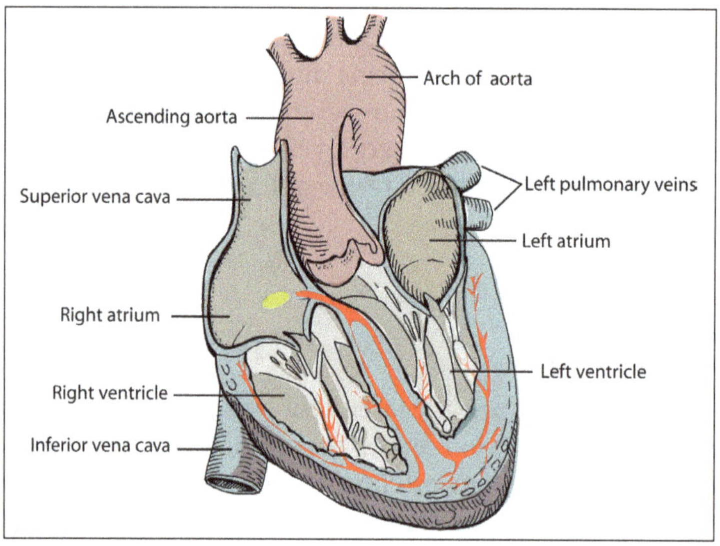

Heart.

The heart is divided into a right and left side. Each side is then divided into two parts called the atrium and the ventricle. So there is a right and left ventricle, and a right and left atrium. A large valve called the atrioventricular valve, or AV valve, is located between the atrium and the ventricle of each side of the heart. The left AV valve is also called the bicuspid valve. The right AV valve is also known as the tricuspid valve. Heart valves have either two or three flaps of tissue, called leaflets, which open and close. Healthy valve leaflets ensure that the blood flows in only one direction through the heart. Closed AV valves prevent blood from flowing backward into the atria when the heart beats.

Two valves prevent blood from rushing back into the respective ventricles. They are the aortic valve and the pulmonary valve. The aortic valve is located between the junction of the left ventricle and the aorta artery.

The pulmonary valve is located on the right side of the heart between the junction of the right ventricle and the pulmonary artery. The best way to understand the operation of the heart is to follow the path that blood takes through the heart and lungs. Unoxygenated blood returns to the heart from the body by way of the cranial and caudal vena cava to the right atrium. Blood then passes through the right AV valve into the right ventricle. From the right ventricle, the blood is pumped into the pulmonary artery, which quickly branches into two parts, each going to a lung. In the alveoli of the lungs, body waste, as carbon dioxide, is exchanged for oxygen.

The blood then returns to the heart (from the lungs) with a fresh supply of oxygen. It goes through the pulmonary vein to the left atrium of the heart. The blood then passes to the left AV valve into the thickwalled left ventricle. From the left ventricle, blood is pumped through the aortic valve into the aorta. The aorta then branches into smaller arteries that carry the oxygenated blood to all parts of the body, including the heart and lungs.

The beat of the heart is controlled by the action of the sinoatrial node, or SA node. The SA node is a group of cells located in the right atrium that sends out electrical signals to make the heart pump. The SA node is the heart's natural pacemaker. These signals travel from the SA node, through the AV node, and then to the remainder of the heart. The SA node also responds to a need for a faster heart rate. For example, if an animal is excited, its body requires greater blood circulation. A healthy SA node responds to this body change and increases the heart rate accordingly.

Anatomy and Physiology of the Vascular System

The blood vessels that carry oxygen-rich blood from the heart to other parts of the body are called arteries. The walls of the large arteries are thick and are composed of an elastic-like tissue. This enables them to withstand the blood pressure created by the heart's beating. Farther away from the heart, arteries branch into smaller arteries called arterioles that then branch into the smallest arteries called capillaries. The smaller arteries contain a large amount of smooth muscle in their walls as opposed to the elastic tissue of the large arteries.

Arterioles have an especially thick layer of smooth muscle in their walls at the junction of the arterioles and capillaries. The purpose of this is to carefully control the amount of blood that each capillary receives. This tension at the ends of the arterioles is very important in maintaining blood pressure throughout the circulatory system. During shock, the arterioles dilate, or relax, allowing a large volume of blood to be contained within the capillary beds.

Capillaries are the blood vessels that connect arteries to veins. They are tiny, thin-walled vessels located in all body tissues. The diameter of capillaries allows blood cells to pass through in single file. The walls of the capillaries act as a semi-permeable membrane allowing the transfer of nutrients, oxygen, and water from the blood to the tissues and waste products from the tissues to the blood. There are also larger tubular connectors of the arterioles to venules. They are located within the capillary beds. These tubules allow more blood to flow through an area. They also help in warming tissues and increasing the return of blood pressure to the heart.

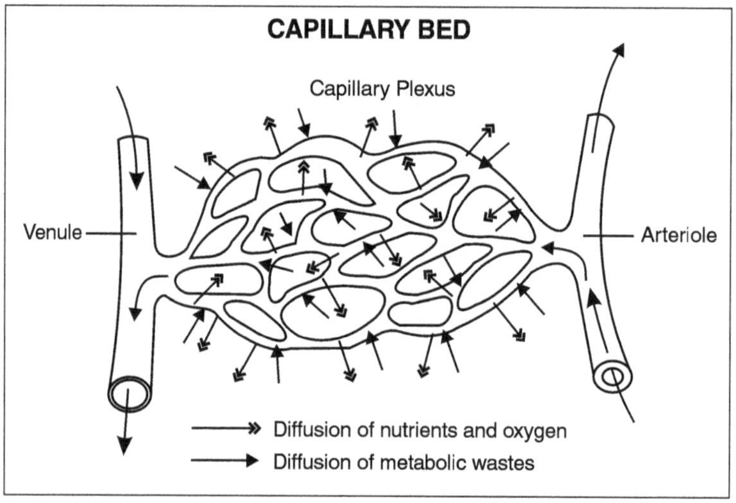

Interaction of molecules flowing in and out of blood at a capillary bed.

Veins are the blood vessels that return blood to the heart from all parts of the body. Capillaries unite to form small veins called venules that then unite to form larger veins closer to the heart. In comparison, veins have thin walls that are collapsible. Veins are much larger than their artery counterparts and contain valves, which aid in the return flow of blood to the heart. The valves are often located at a point where several smaller veins unite to form larger veins to prevent the reverse flow of blood. The valves also allow contraction of muscles and movement of body parts. This also aids the return flow of blood to the heart, if blood pressure in the veins is low.

Anatomy and Physiology of Systems within the Circulatory System

The part of the circulatory system that transports blood through the lungs, where it is oxygenated, is called pulmonary circulation. Pulmonary circulation is the movement of blood from the heart, to the lungs, and back to the heart again. The veins return waste-laden blood to the heart, entering the right atrium through two large veins called vena cavae. The right atrium fills with the waste-laden blood and then contracts, pushing the blood through a one-way valve into the right ventricle. The right ventricle fills and then contracts, pushing the blood into the pulmonary artery, which leads to the lungs. The pulmonary artery divides into two branches after leaving the heart, with each branch going to a lung.

The exchange of carbon dioxide and oxygen takes place in the lungs. The pulmonary artery branches subdivide into arterioles that finally subdivide into many capillary beds in the lungs. These capillary beds are closely associated with alveoli (the smallest air passages of the lungs).

A minimum amount of tissue separates the blood in the capillary beds from the alveoli. This allows carbon dioxide (the gaseous waste of animal metabolism) to be exchanged for oxygen from the air. Oxygenated arterial blood is bright red in color, and unoxygenated venous blood is dark or brownish red.

The fresh, oxygen-rich blood enters the pulmonary veins and then returns to the heart, re-entering through the left atrium. The oxygen-rich blood then passes through a one-way valve into the left ventricle where it will exit the heart through the aorta. The left ventricle's contraction forces the blood into the aorta and the blood begins its journey throughout the body.

The pressure required for pulmonary circulation is much less than that required for systemic, or body circulation. Therefore, the muscle mass developed in the right ventricle is much less than that of the left ventricle.

The heart and the vessels that move oxygenated arterial blood to all parts of the body and return the unoxygenated venous blood to the heart are referred to as the systemic circulatory system. The blood vessels (arteries, veins, and capillaries) are responsible for the delivery of oxygen and nutrients to the tissue. Oxygen-rich blood enters the blood vessels through the heart's main artery called the aorta. The forceful contraction of the heart's left ventricle forces the blood into the aorta, which then branches into many smaller arteries that run throughout the body. The inside layer of an artery is very smooth, allowing the blood to flow quickly. The outside layer of an artery is very strong, allowing the blood to flow forcefully. The oxygen-rich blood enters the capillaries where the oxygen and nutrients are released. The waste products are collected, and the waste-laden blood flows into the veins to return to the heart where pulmonary circulation will allow the exchange of gases in the lungs.

During systemic circulation, blood passes through the kidneys. This phase of systemic circulation is known as renal circulation. During this phase, the kidneys filter much of the waste from the blood. Blood also passes through the small intestine during systemic circulation. This phase is known as portal circulation. During this phase, the blood from the small intestine collects in the hepatic portal vein, which passes through the liver. The liver filters sugars from the blood and stores them for later use. The systemic circulatory system is complex and its functions vary with different tissue requirements (i.e., exercise versus digestion). The subystems within the systemic circulatory system each serve a particular region of the body. The major divisions of the systemic circulatory system are served by the aorta, thoracic aorta, abdominal aorta, and arteries that serve the hepatic portal system. The anterior vena cava and the posterior vena cava are the major veins of the systemic circulatory system.

Anatomy and Physiology of the Total Circulatory System

The largest artery leaving the left ventricle of the heart carrying oxygenated blood is called the aorta. The right and left coronary arteries branch from the aorta. A heart attack often involves a clot in the coronary arteries or their branches. The coronary arteries supply fresh blood to the heart muscle itself. The blood is returned quickly to the heart by the coronary veins. The next branch of the aorta following the coronary branches is the brachiocephalic trunk. The carotid arteries branch off the brachiocephalic artery. They extend toward the head of the animal on either side of the neck. These arteries supply oxygenated blood to the neck and head region. This blood is returned by the jugular vein, which is usually prominent on the sides of the neck of animals. The right and left brachial arteries that also branch off the brachiocephalic trunk supply oxygenated blood to the shoulders and forelegs.

When leaving the heart, the aorta travels upward to the thoracic vertebrae where it becomes the thoracic aorta. It extends to the rear of the animal and becomes the abdominal aorta after it passes the diaphragm. Around the last lumbar vertebrae, the aorta branches into two external and internal iliac arteries.

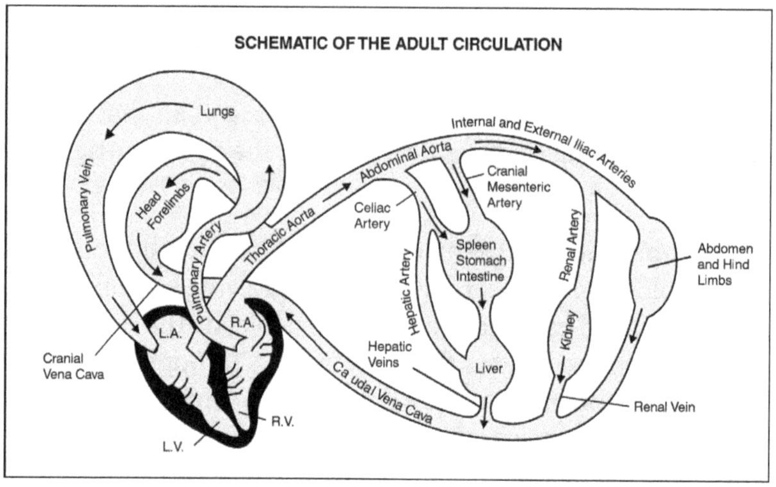

The thoracic aorta branches in the thoracic area. These branches provide oxygenated blood to the lungs (called the bronchial arteries), esophagus, ribs, and diaphragm. Immediately after the aorta passes through the diaphragm, the celiac artery branches into the gastric, splenic, and hepatic arteries. These arteries supply oxygenated blood to the stomach, spleen, and liver, respectively. Then, the abdominal aorta branches into the cranial and caudal mesenteric arteries. These arteries supply oxygenated blood to the small and large intestines.

The next major branch of the abdominal aorta consists of the renal arteries that supply oxygenated blood to the kidneys. The function of the renal arteries is not only to supply blood to the kidneys, but also to carry large volumes of blood to the kidneys for filtration and purification. The renal arteries also give rise to arteries that supply blood to the testicles of males and parts of the reproductive system of females. Those arteries are the internal spermatic arteries and the uteroovarian arteries, respectively.

The abdominal aorta ends in the paired lumbar veins and the internal and external iliac arteries. The internal iliac arteries provide blood to the pelvic and hip region. The external iliac arteries branch into the femoral arteries and their branches, which provide oxygenated blood to the hindlegs. Veins normally accompany arteries and many times have similar names. The veins are always larger than the arteries and sometimes more numerous. Most all veins eventually empty into the caudal or cranial vena cava in returning unoxygenated blood to the right atrium of the heart. Sometimes veins are closer to the skin surface than are the arteries and, therefore, are more visible.

The cranial vena cavae veins are responsible for draining the unoxygenated blood from the head, neck, forelegs, and part of the thorax, or chest cavity. Some major veins in this area are the jugular veins, brachial veins, internal thoracic veins, and the vertebral veins. The caudal vena cava collects blood from the iliac, lumbar, renal, and adrenal veins to return it to the right atrium of the heart.

The part of the systemic system that carries venous blood from the stomach, pancreas, small intestine, and spleen to the liver is called the hepatic portal system. In the liver, the blood passes through a second capillary bed. The gastric vein from the stomach, the splenic vein from the spleen, the pancreatic veins from the pancreas, and the mesenteric veins from the intestines empty into the portal vein. The portal vein carries blood to the liver from these areas to be purified and for nutrients to be stored for future use. The portal vein branches into smaller venules and finally into the capillary beds of the liver. These capillaries join into venules that empty into the hepatic vein, which carries unoxygenated blood to the caudal vena cava.

SYSTEMIC BLOOD SYSTEM OF THE COW

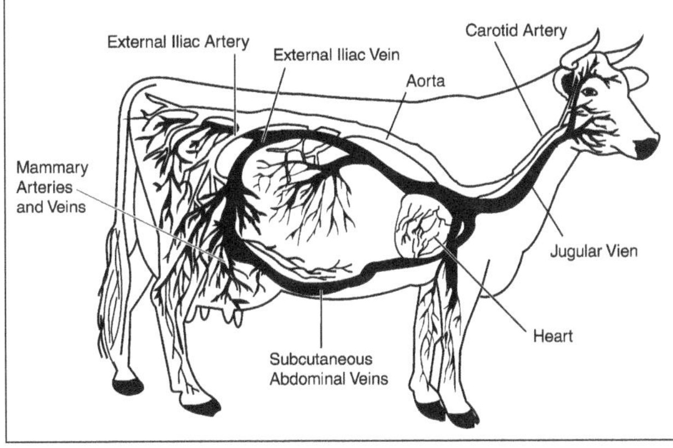

Anatomy and Physiology of the Lymphatic System

The lymphatic system is an accessory to the circulatory system. The thin walls of capillaries allow fluid and gases to escape into spaces between body tissue cells where lymphoid tissue containing lymphocytes are located. Many of these fluids do not reenter capillaries, but are picked up by very thin-walled structures called lymph vessels. The lymph vessels start in the tissue spaces and form larger ducts as they pass through lymph glands or nodes. They finally empty into large blood veins such as the cranial vena cava. Lymph glands are responsible for filtering foreign substances from the lymph, thus preventing them from entering the blood system.

Lymph glands, or nodes, are scattered among the lymph vessels and serve as the body's first defense against infection. They produce lymphocytes and antibodies against diseases. Each lymph gland consists of a cortex (outer covering) and a medulla. Other than having large numbers of lymphocytes in the medulla, the lymph glands produce plasma cells that are larger than lymphocytes. Plasma cells are an important source of antibody production in animals. Each lymph gland has its own blood supply and venous drainage.

FLOW OF BLOOD AND LYMPH WITHIN TISSUE

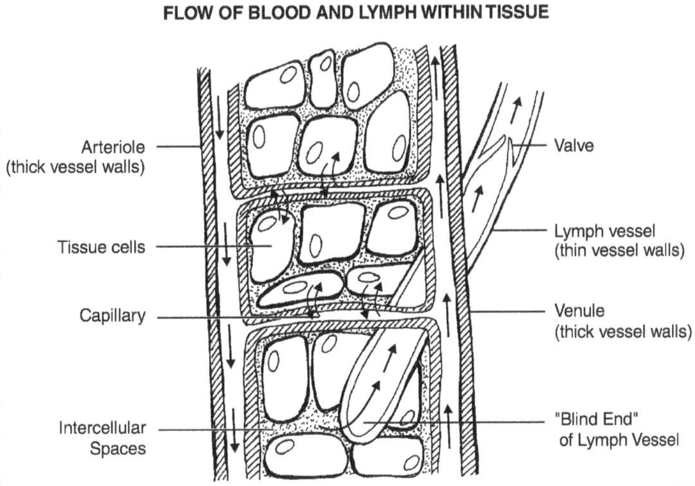

When an infection occurs in a specific location, the lymph glands in that area will enlarge to combat the infection. Sometimes the first lymph node closest to the infection is unable to eliminate the infection. Then, other lymph glands along the lymphatic system will attempt to battle the infection. This is particularly critical in the case of cancer. Through the lymphatic system, cancer can be spread from the point of its beginning and all parts of the body. Lymph glands are usually named in relation to their location in the body. For example, the mandibular lymph gland is located just under the mandible, or jaw, in the angle formed by the mandible and the neck.

Reproductive System

Animal reproductive system is any of the organ systems by which animals reproduce. The role of reproduction is to provide for the continued existence of a species; it is the process by which living organisms duplicate themselves. Animals compete with other individuals in the environment to maintain themselves for a period of time sufficient to enable them to produce tissue nonessential to their own survival, but indispensable to the maintenance of the species. The additional tissue, reproductive tissue, usually becomes separated from the individual to form a new, independent organism.

Unlike most other organ systems, the reproductive systems of higher animals have not generally become more complex than those of lower forms. Asexual reproduction (i.e., reproduction not involving the union of gametes), however, occurs only in the invertebrates, in which it is common, occurring in animals as highly evolved as the sea squirts, which are closely related to the vertebrates. Temporary gonads are common among lower animals; in higher animals, however, gonads are permanent organs. Hermaphroditism, in which one individual contains functional reproductive organs of both sexes, is common among lower invertebrates; yet separate sexes occur in such primitive animals as sponges, and hermaphroditism occurs in animals more highly evolved—e.g., the lower fishes. Gonads located on or near the animal surface are common in the lowest invertebrates, but in higher animals they tend to be more deeply situated and often involve intricate duct systems. In echinoderms, which are among the highest invertebrates, the gonads hang directly into the sea and spill their gametes into the water. In protochordates, gametes are released into a stream of respiratory water that passes directly into the sea. Duct systems of the invertebrate flatworms (Platyhelminthes) are relatively complex, and those of specialized arthropods (e.g., insects, spiders, crabs) are more complex than those of any vertebrate. Copulatory organs occur in flatworms, but copulatory organs are not ubiquitous among vertebrates other than reptiles and mammals. The trend toward fewer eggs and increased parental care in higher animals may account for the relative lack of complexity in the reproductive systems of some advanced forms. Whereas trends toward increasing structural complexity have often been reversed during evolution, reproductive behaviour patterns in many phylogenetic (i.e., evolutionary) lines have become more complicated in order to enhance the opportunity for fertilization of eggs and maximum survival of offspring.

A direct relationship exists between behaviour and the functional state of gonads. Reproductive behaviour induced principally but not exclusively by organic substances called hormones promotes the union of sperm (spermatozoa) and eggs, as well as any parental care accorded the young. There are a number of reasons why behaviour must be synchronized with gonadal activity. Chief among these are the following:

- Individuals of a species must congregate at the time the gonads contain mature gametes. This often entails migration, and some members of all major vertebrate groups migrate long distances to gather at spawning grounds or rookeries.
- Individuals with gametes ready to be shed must recognize members of the opposite sex. Recognition is sometimes by external appearance or by chemical substances (pheromones), but sex-linked behaviour is often the only signal.
- Geographical territories frequently must be established and aggressively defended.
- The building of nests, however simple, is essential reproductive behaviour in many species.
- When fertilization of aquatic forms is external, sperm and eggs must be discharged at approximately the same time into the water, since gametes may be quickly dispersed by currents. Courtship, often involving highly intricate behaviour patterns, serves to release the gametes of both mating individuals simultaneously.
- When fertilization is internal, willingness of the female to mate is often essential. Female mammals not in a state of willingness to mate not only will not mate but may injure or even kill an aggressive male. The unwillingness of a female mammal to mate when mature eggs are not present prevents loss of sperm needed to preserve the species.
- Parental care of fertilized eggs by one parent or the other has evolved in many species. Parental behaviour includes fanning the water or air around the eggs, thereby maintaining appropriate temperature and oxygen levels; secretion of oxygen from a parent's gills; transport of eggs on or in the parental body (including the mouth of some male parents); and brooding, or incubation, of eggs.
- Some species extend parental care into the postnatal period, feeding and protecting the offspring. Such behaviour patterns are adaptations for survival and thus are essential; all are induced by the nervous and endocrine systems and are typically cyclical, because gonadal activity is cyclical.

Reproductive Systems of Invertebrates

Gonads, Associated Structures and Products

Although asexual reproduction occurs in many invertebrate species, most reproduce sexually. The basic unit of sexual reproduction is a gamete (sperm or egg), produced by specialized tissues or organs called gonads. Sexual reproduction does not necessarily imply copulation or even a union of gametes. As might be expected of such a large and diverse group as the invertebrates, many variations have evolved to ensure survival of species. In many lower invertebrates, gonads are temporary organs; in higher forms, however, they are permanent. Some invertebrates have coexistent female and male gonads; in others the same gonad produces both sperm and eggs. Animals in which both sperm and eggs are produced by the same individual (hermaphroditism) are termed monoecious. In dioecious species, the sexes are separate. Generally, the male gonads ripen first in hermaphroditic animals (protandry); this tends to ensure cross-fertilization. Self-fertilization is normal, however, in many species, and some species undergo sex reversal.

Sponges, Coelenterates, Flatworms and Aschelminths

Sponges are at a cellular level of organization and thus do not have organs or even well-developed tissues; nevertheless, they produce sperm and eggs and also reproduce asexually. Some species of sponge are monoecious, others are dioecious. Sperm and eggs are formed by aggregations of cells called amoebocytes in the body wall; these are not considered gonads because of their origin and transitory nature.

In hydrozoan coelenterates, temporary gonads are formed by groups of cells in either the epidermis (outer cell layer) or gastrodermis (gut lining), depending on the species; scyphozoan and anthozoan coelenterates generally have gonads in the gastrodermis. The origin and development of gonads in coelenterates, particularly freshwater species, are often associated with the seasons. Freshwater hydrozoans, for example, reproduce asexually until the onset of cold weather, which stimulates them to form testes and ovaries. Colonial hydrozoans asexually produce individuals known as polyps. Polyps, in turn, give rise to free swimming stages (medusae), in which gonads develop. The body organization of sponges and coelenterates is such that most of their cells are in intimate contact with the environment; consequently, gametes are shed into the water, and no ducts are necessary to convey them to the outside.

In contrast to sponges and coelenterates, platyhelminths generally have well-developed organ systems of a permanent nature and, in addition, have evolved secondary reproductive structures to convey sex products. One exception is the acoels, a group of primitive turbellarians; they lack permanent gonads, and germinal cells develop from amoebocytes in much the same manner as in sponges. The majority of flatworms, however, are monoecious, the primary sex organs consisting of one or more ovaries and testes. The tube from the ovary to the outside is called the oviduct; it often has an outpocketing (seminal receptacle) for the storage of sperm received during copulation. In many species the oviduct receives a duct from yolk (vitelline) glands, whose cells nourish the fertilized egg. Beyond the entrance of the duct from the yolk glands the oviduct may be modified to secrete a protective capsule around the egg before it is discharged to the outside. The male organs consist of testes, from which extend numerous tubules (vasa efferentia) that unite to form a sperm duct (vas deferens); the latter becomes an ejaculatory duct through which sperm are released to the outside. The sperm duct may exhibit expanded areas that store sperm (seminal vesicles), and it may be surrounded by prostatic cells that contribute to the seminal fluid. The sperm duct eventually passes through a copulatory organ. The same basic structural pattern, somewhat modified, is found in higher invertebrates.

Aschelminthes (roundworms) are mostly dioecious; frequently there are external differences between males and females (sexual dimorphism). The males are generally smaller and often have copulatory spicules. Nematodes have relatively simple reproductive organs, a tubular testis or ovary being located at the end of a twisted tube. The portion of the female tract nearest the ovary forms a uterus for temporary storage of fertilized eggs. Some species lay eggs, but others retain the egg in the uterus until the larva hatches. The sperm are released into a cavity called the cloaca. A number of free-living nematodes are capable of sex reversal—if the sex ratio in a given population is not optimal or if environmental conditions are not ideal, the ratio of males to females can be altered.

This sometimes results in intersexes; i.e., females with some male characteristics. Hermaphroditism occurs in nematodes, and self-fertilization in such species is common.

Annelids and Mollusks

Annelids have a well-developed body cavity (coelom), a part of the lining of which gives rise to gonads. In some annelids, gonads occur in several successive body segments. This is true, for example, in polychaetes, most of which are dioecious. Testes and ovaries usually develop, though not invariably, in many body segments; and the sperm and eggs, often in enormous numbers, are stored in the coelom. Fertilization is external. In oligochaetes (all of which are monoecious) on the other hand, the gonads develop in a few specific segments. Sperm are stored in a seminal vesicle and eggs in an egg sac, rather than in the coelom. A portion of the peritoneum, the membrane lining the coelom, becomes a saclike seminal receptacle that stores sperm received from the mate. The earthworm, Lumbricus terrestris, is an example of a specialized annelid reproductive system. Female organs consist of a pair of ovaries in segment 13; a pair of oviducts that open via a ciliated funnel (i.e., with hairlike structures) into segment 13 but open to the exterior in segment 14; an egg sac near each funnel; and a pair of seminal receptacles in segment 9 and another in segment 10. Male organs consist of two pairs of minute testes in segments 10 and 11, each associated with a ciliated sperm funnel leading to a tiny duct, the vas efferens. The two ducts on each side lead to a vas deferens that opens in segment 15. Testes and funnels are contained within two of three pairs of large seminal vesicles that occupy six body segments. Leeches (Hirudinea), also monoecious, have one pair of ovaries and a segmentally arranged series of testes with duct systems basically similar to those of earthworms.

Although mollusks have a close evolutionary kinship to annelids, they have reduced or lost many structures characteristic of segmented worms. The coelom persists only as three regional cavities: gonadal, nephridial (kidney), and pericardial (heart). In ancestral forms these were interconnected so that gametes from the gonad passed through the pericardial cavity, the nephridial cavity, then to the outside through a nephridial pore. The various groups of mollusks have tended to modify this arrangement, with the result that gonads have their own pore; among amphineurans, for example, the sexes are usually separate, and there is one gonad with an associated pore. Gastropods show considerable variability, but generally one gonad (ovary, testis, or ovotestis—a structure combining the functional gonads of both sexes) is located in the visceral hump and connected to the outside by a remnant of the right kidney. In hermaphroditic forms, one duct carries sperm as well as eggs. The gonadal ducts of gastropod females often secrete a protective capsule around the fertilized eggs; in males, the terminal portion of the duct is sometimes contained in a copulatory organ. Pelecypods may be either monoecious or dioecious, but the gonads are usually paired. In mussels and oysters, the gonads open through the nephridial pore, but in clams the reproductive system opens independently. The cephalopods are all dioecious. The single testis or ovary releases its products into the pericardial cavity and this, in turn, leads to a gonopore, the external opening. The oviduct of the squid is terminally modified to form a shell gland. The male system is more complex—the gonoduct leads into a seminal vesicle where a complicated torpedo-shaped sperm case (spermatophore) is secreted and contains the sperm. Spermatophores are then stored in a special structure (Needham's sac) until copulation occurs.

A remarkable characteristic of some mollusks is the ability to alter their sex. Some species are clearly dioecious; however, among the monoecious species there is considerable variability in their hermaphroditic condition. In some species, male and female gonads, although in the same individual, are independent functionally and structurally. In others, an ovotestis produces both sperm and eggs. Oysters display a third condition; young oysters have a tendency toward maleness, but, if water temperature or food availability is altered, some individuals develop into females. Later, a reversal to the male condition may occur. The sexual makeup of an entire oyster population also has a seasonal aspect; in harmony with the group, an individual may undergo several alterations in the course of a year. A similar phenomenon, called consecutive sexuality, occurs in limpets. These gastropods stack themselves in piles, with the younger animals on top. The animals on top are males with well-developed testes and copulatory organs; those in the middle are hermaphroditic; those on the bottom are females, having lost the testes and copulatory organ (penis) by degeneration. A decrease in the number of females in a stack induces males to assume female characteristics, but the transition is retarded when an excess of females is present. The degree of maleness or femaleness is probably controlled in part by environmental and internal factors.

Arthropods

The phylum Arthropoda includes a vast number of organisms of great diversity. Most arthropods are dioecious, but many are hermaphroditic, and some reproduce parthenogenetically (i.e., without fertilization). The primary reproductive organs are much the same as in other higher invertebrates, but the secondary structures are often greatly modified. Such modifications depend on whether fertilization is internal or external, whether the egg or zygote (i.e., the fertilized egg) is retained or immediately released, and whether eggs are provided some means of protection after they have left the body of the female. The mandibulate arthropods (e.g., crustaceans, insects) include more species than any other group and have invaded most habitats, a fact reflected in their reproductive processes.

Crustaceans (e.g., crabs, crayfish, and barnacles) are for the most part dioecious. The primary reproductive organs generally consist of paired gonads that open through paired ventral (bottom side) gonopores. Females often have a seminal receptacle (spermatheca) in the form of an outpocketing of the lower part of the female tract or as an invagination (inpocketing) of the body near the gonopore. Males have appendages modified for clasping the female during copulation or for guiding sperm. A number of groups have members that reproduce parthenogenetically. Branchiopods (e.g., water fleas, fairy shrimp) have simple paired gonads. The female gonopore often opens dorsally (on the back side) into a brood chamber; the male gonopore opens near the anus. Males have appendages for clasping females during copulation. Ostracods, or seed shrimp, have paired tubular gonads. The eggs may be brooded by the female, or they may be released into the water via a gonoduct and gonopore. The terminal portion of the male gonoduct is enclosed in a single or paired penis. Many species reproduce parthenogenetically. Some experts contend that this is the only method employed, even though functional males may be present in the population. Copepods (e.g., Cyclops) have paired ovaries and an unpaired testis. The terminal portion of the oviduct constitutes an ovisac for storage of eggs. The male deposits sperm in a spermatophore that is transferred to the female. Sexual dimorphism is particularly evident among parasitic copepods. Frequently, parasitic females can hardly be recognized as copepods except for the distinctive ovisacs. Males, on the other hand, are free-living and are recognizable as copepods.

The hermaphroditic Cirripedia (e.g., barnacles) are among the exceptions to the generalization that crustaceans are dioecious. It has been suggested that hermaphroditism in barnacles is an adaptation to their sessile, or stationary, existence, but cross-fertilization is more common than self-fertilization. The ovaries lie either in the base or in the stalk of the animal, and the female gonopore is near the base of the first pair of middle appendages (cirri). The testes empty into a seminal vesicle through a series of ducts; from the vesicle extends a long sperm duct within a penis that may be extended to deposit sperm in the mantle cavity of an adjacent barnacle. The terminal portion of the oviduct secretes a substance that forms a kind of ovisac within the mantle cavity, where fertilized eggs undergo early development. Although most barnacles are hermaphrodites, some display a peculiar adaptation in that they contain parasitic dwarf or accessory males. Dwarf males are much smaller than the host barnacle in which they live and are degenerate, except for the testes. In some species they live in the mantle cavity of hermaphroditic forms and produce accessory sperm; in other species only the female organs persist in the host animal and the accessory male is a necessity.

Amphipods and isopods (e.g., pill bugs, sow bugs), like most crustaceans, are dioecious and have paired gonads. Females of both groups have a ventral brood chamber (marsupium) formed by a series of medially directed (i.e., toward the body midline) plates (oostegites) in the region of the thorax, the region between head and abdomen. Many isopods are parasitic and have developed unusual sex-related activities. Certain species are parasitic on other crustaceans. After a series of molts (i.e., shedding of the body covering) a parasitic larval (immature) isopod attaches to the shell of a crab. If it is the only larva to do so, it increases in size and develops into an adult female. If another larva subsequently attaches, the new arrival becomes a male. It has been demonstrated that the testes of the functioning male larva will change to ovaries if the larva is removed to a new, uninfected host. Thus, the larvae of these species apparently are intersexual and can develop into either sex. This phenomenon, reminiscent of that in mollusks, demonstrates the way in which similar adaptations have evolved in diverse groups of organisms.

The gonads of crabs and lobsters are paired, as are the gonopores. The females of many species have external seminal receptacles on the ventral part of the thorax; those of other species have internal receptacles in the same region. In some species, seminal receptacles are absent, and the male simply attaches a spermatophore to the female. Thus, males either have appendages (gonopods) by which sperm are inserted in the body of the female or produce spermatophores for sperm transfer. The sexual dimorphism of many decapods can be altered by parasitism. An example of this is the crab that is parasitized by a barnacle. A barnacle infection in male crabs induces the secondary sex characters of the crab to resemble those of a female; however, masculinization does not occur in parasitized females. At each molt a parasitized male crab increasingly resembles a female even though the testes may be completely unaffected. Feminization results from a hormonal alteration of the parasitized crab.

Insects are rarely hermaphroditic, but many species reproduce parthenogenetically (without fertilization). The insect ovary is composed of clusters of tubules (ovarioles) with no lumen, or cavity. The upper portion of each ovariole gives rise to oocytes (immature eggs) that mature and are nourished by yolk from the lower portion. The oviduct leads to a genital chamber (copulatory bursa, or vagina), with which are often associated accessory glands and a seminal receptacle. Some accessory glands form secretions by which eggs become attached to a hard surface; others secrete a protective envelope around the egg. The eighth and ninth body segments are often modified for egg-laying. The paired testes consist of a series of seminal tubules that form primary spermatogonia (immature spermatozoa) at their upper ends. As the spermatogonia mature a covering is secreted around them. Eventually they enter a storage area (seminal vesicle). The terminal portion of the male system is an ejaculatory duct that passes through a copulatory organ. A pair of accessory glands, often associated with the ejaculatory duct, contributes to the semen (fluid containing sperm) or participates in spermatophore formation. The ninth body segment and sometimes the tenth bear appendages for sperm transfer. Scorpions and spiders have tubular or saclike gonads; the female system is equipped to receive and store sperm, and, in some species, the female retains the eggs long after fertilization has occurred. Male spiders may have a cluster of accessory glands associated with the terminal portion of the reproductive system for the manufacture of spermatophores, or they may have expanded seminal vesicles for the retention of sperm until copulation takes place. Often specific appendages are adapted for sperm transfer.

Echinoderms and Protochordates

Echinoderms (e.g., sea urchins), hemichordates (including acornworms), urochordates (e.g., sea squirts), and cephalochordates (amphioxus) are restricted to a marine habitat. As with many other marine animals, their gametes are shed into the water. In echinoderms, the gonads are generally suspended from the arms directly into the sea; with few exceptions, the sexes are separate. Female starfishes have been known to release as many as 2,500,000 eggs in two hours; 200,000,000 may be shed in a season. Males produce many times that number of sperm. Acornworms reproduce only sexually, and the sexes are generally separate. The gonads lie on each side of the gut as a paired series of simple or lobed sacs. Each opens to the exterior, either directly or via a short duct. The eggs, when shed, are in coiled mucous masses, each of which contains 2,500 to 3,000 eggs.

In urochordates and cephalochordates the gonads develop in the wall of a cavity (atrium) that receives respiratory water after it passes over the gills. Gametes are released into the cavity, and then carried into the sea by the water flowing from the cavity. Most urochordates are hermaphroditic. One ovary and one testis may lay side by side, each with its own duct to the atrium; some species have many pairs of ovaries and testes. The eggs develop in so-called ovarian follicles consisting of two layers of cells, as in many vertebrates. The inner layer remains with the ovulated, or shed, egg, and the cells become filled with air spaces, which apparently help the eggs to float. Amphioxus, the highest animals lacking vertebral columns, is dioecious. They have 24 or more pairs of ovaries or testes lacking ducts. When ripe, the gonads rupture, spilling their gametes directly into the atrium.

Mechanisms that Aid in the Union of Gametes

Sponges, Coelenterates, Flatworms, and Aschelminthes

The processes of sperm transfer and fertilization have been documented for only a few species of sponges. Flagellated (i.e., bearing a whiplike strand) sperm are released from the male gonad and swept out of the body and into the water by way of an elaborate system of canals. A sperm that enters another sponge, or the one from which it was released, is captured by a flagellated collar cell (choanocyte). The choanocyte completely engulfs the sperm, loses its collar and flagellum (or "whip"), and migrates to deeper tissue where the egg has matured. The choanocyte containing the sperm cell fuses with the egg, thus achieving fertilization. In freshwater coelenterates, sperm are also released into the water and carried by currents to another individual. Unlike the mechanism in sponges, however, coelenterate eggs arise in the epidermis, or surface tissue, and are exposed to sperm that may be nearby in the water; thus, no intermediate transport cell is needed. Many species of marine coelenterates expel both sperm and eggs into the water, and fertilization takes place there. Some medusoid coelenterates (jellyfish), however, offer some protection to the egg. After leaving the gonad, the egg becomes temporarily lodged in the epidermis on the underside of the organism, where fertilization and early development occur.

In all flatworms, fertilization is internal. Among species with no female duct, sperm are injected, and fertilization occurs in the inner layer of tissue. Most flatworms, however, have an elaborate system of male and female ducts. Generally, the male gonoduct passes through a penis-like organ, and sperm are transferred during copulation. In parasitic species, which often cannot find a mate, self-fertilization serves as the means for reproduction. Sperm and ova unite in the oviduct, which then secretes yolk around the zygote.

Male nematodes (roundworms) are usually equipped with a pair of copulatory structures (spicules) that guide the sperm during copulation. The posterior end of some males also exhibits a lateral (sideward) expansion (copulatory bursa) that clasps the female during copulation. Other males loop their tail around the female in the region of her gonopore. Unlike many other aschelminthes, nematodes have sperm cells that are amoeboid; i.e., their cell contents seem to flow. Some male rotifers have a copulatory organ.

Annelids and Mollusks

In some species of annelid polychaetes (marine worms) reproductive activity is synchronized with lunar cycles. At breeding time the body of both sexes differentiates into two regions, an anterior atoke and a posterior epitoke, in which gonads develop. When the moon is in a specific phase, the epitoke separates from the rest of the body and swims to the surface. The female epitoke apparently stimulates the male epitoke to release sperm, and sperm release, in turn, evokes expulsion of eggs. Fertilization is external. So well-coordinated is this phenomenon that tremendous numbers of epitokes appear on the surface at about the same time.

Sexually mature oligochaetes have a clitellum, which is a modification of a section of the body wall consisting of a glandular, saddlelike thickening near the gonopores. During copulation, the clitellum secretes mucus that keeps the worms paired while sperm are being exchanged. Following copulation, the clitellum secretes substance for a cocoon, which encircles the worm and into which eggs and sperm are deposited. The worm then manipulates the cocoon until it slips off over the head. Thereupon, the ends of the cocoon become sealed, and fertilization and development take place inside. Many leeches also form a cocoon; but the males of some species have a penis that can be inserted into the female gonopore. In other leeches, a spermatophore is thrust into the body of the mate during copulation.

Union of gametes among mollusks is affected in a number of ways. Marine pelecypods synchronously discharge sperm and eggs into the sea; some freshwater clams are apparently self-fertilizing. One of the more unusual types of reproductive diversity occurs in marine gastropods of the family Scalidae that produce two kinds of sperm cells. A large sperm with a degenerate nucleus acts as a transport cell for carrying numerous small fertilizing sperm through the water and into the oviduct of another individual. Cephalopod males have modified arms for the transfer of spermatophores. The right or left fourth arm of the squid, for example, is so modified. Following an often elaborate courtship, the male squid uses the modified appendage to remove spermatophores from their storage place in his body and place them in the mantle cavity of the female. A cementing substance, which is released from the spermatophore, firmly attaches the spermatophore to the female's body near the oviduct. In some species, the male loses the arm. Manipulation of the eggs by the female's arms may also occur.

Some unusual behaviour patterns have evolved in conjunction with sperm transfer in mollusks. Prior to copulation of certain land snails, a dart composed of calcium carbonate is propelled forcefully from the gonopore of each of the mating individuals and lodges in the viscera of the mate. Even though the snails have assumed a mating posture, sperm transfer cannot occur until each snail has been stimulated by a dart.

Arthropods

Arthropods are as varied as mollusks in their methods of effecting union of sperm and eggs. They have relatively few devices for sperm transfer, but many display a high degree of behavioral complexity.

The male and female scorpion participates in a courtship ritual involving complicated manoeuvres. In some species the male produces spermatophores that are anchored to the ground. In the course of the ritual dance the female is positioned over the spermatophore. The male then presses her down until the sperm packet is forced into her genital chamber, where it becomes attached by means of small hooks. Thus, ultimately, fertilization takes place internally.

Among some spiders the male's pedipalp, a grasping or crushing appendage, contains a bulb and an extensible, coiled structure (embolus). As mating begins, the male dips the pedipalp into semen from his gonopore. The embolus is then placed in the female gonopore, and the sperm are transferred to her seminal receptacle. The female deposits the sperm along with her eggs into a silken cocoon, which she attaches to her body or to an object such as a rock or a leaf.

Sperm transfer in copepods, isopods, and many decapods, often preceded by courtship, is affected by modified appendages, gonopods, or spermatophores. Copepods clasp the female with their antennae while placing a spermatophore at the opening of the seminal receptacle. In some decapods fertilization occurs as eggs are being released into the water.

Fertilization among insects is always internal; there is much variation in the manner in which sperm are transferred to the female. Males of some species form spermatophores that are deposited in a copulatory bursa (vagina) of the female; the wall of the spermatophore breaks down, and the sperm swim to the seminal receptacle. In other species, sperm are introduced directly into the seminal receptacle by an intromittent organ. In still others, sperm, but no spermatophores, are deposited in the copulatory bursa and migrate to the seminal receptacle. In all instances, sperm are retained in the seminal receptacle until after fertilization. An exception to the usual route of sperm transfer occurs in insects that inject sperm into the female's hemocoel (i.e., the space between the body organs). The sperm then migrate to the ovary and oviduct and unite with eggs before the eggshell is formed.

Parthenogenesis

Most frequently, parthenogenesis is the development of a new individual from an unfertilized gamete. Often referred to as unisexual reproduction, it has been observed in almost every major invertebrate group, with the exception of protochordates (including hemichordates), and frequently occurs alternately with bisexual reproduction (reproduction by union of gametes). Some species, in which males are completely unknown, apparently reproduce only by parthenogenesis. Species that alternate between parthenogenesis and bisexual reproduction (heterogenetic species) often do so in response to changes in population density, food availability, or other environmental conditions.

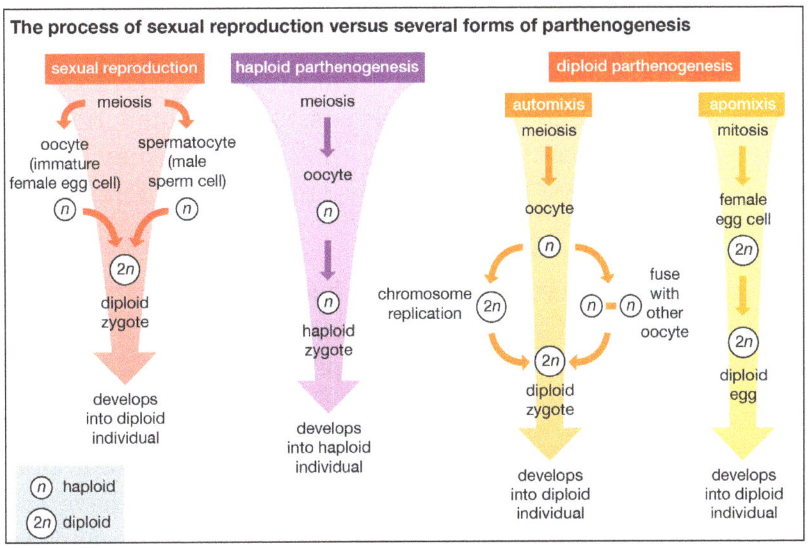

Sexual reproduction; parthenogenesis: The process of sexual reproduction and several forms of parthenogenesis.

The best known examples of parthenogenetic reproduction are found among rotifers. Males are completely unknown in some genera; in others, they appear in the population only for brief periods and more or less seasonally. Females are the dominant form or are the only sex present in a population throughout most of the year. Because no reductional division (meiosis) occurs in the course of egg maturation, the eggs are diploid—thatis, they have the full number of chromosomes; they give rise to new diploid individuals with no chromosomal contribution from a male gamete (diploid parthenogenesis). Even if males were present, sperm could not fertilize the eggs because the latter are already diploid. Under conditions of environmental stress such as seasonal changes, some females form eggs that undergo reductional division, resulting in eggs with the haploid number of chromosomes; such eggs must be fertilized by a male gamete to produce a new female. When the new individual matures, it will probably reproduce parthenogenetically. If, however, there are no males in the population, the haploid eggs can develop into haploid males (haploid parthenogenesis), which then participate in bisexual reproduction. Bisexually produced eggs are often referred to as winter eggs since they have a thick covering that protects the embryo during adverse environmental conditions. Summer eggs, produced parthenogenetically, are thin shelled. Bisexual reproduction occurs, therefore, only often enough to ensure survival of the species.

Nematodes, especially free-living species such as some dioecious soil nematodes, exhibit a type of parthenogenesis known as gynogenesis. In this type of reproduction, the sperm produced by males do not unite with the haploid female egg but merely activate it to begin development. The result is haploid females.

Parthenogenesis, which apparently occurs only rarely in the annelids and mollusks, is found more frequently among the arthropods. The cladocerans (e.g., water fleas), for example, have a reproductive cycle much like that of rotifers—so long as environmental conditions are optimal and food is plentiful, females produce other females by diploid parthenogenesis. When conditions become adverse, males begin to appear in the population, and bisexual reproduction follows. The precise trigger for the appearance of males is not yet known. Fertilized eggs, covered with a highly resistant case, enter a resting stage (ephippium) and can withstand severe temperatures and drying out. The return of favourable conditions leads to the emergence of females that reproduce parthenogenetically. The ability to form a resting stage regulates population density. Whenever the food supply becomes short because of overpopulation by parthenogenetic females, bisexual reproduction is induced, and a dormant stage ensues. During periods of food shortage, the excess females die from lack of food, but the ephippia remain to restore the population.

Insects provide numerous examples of parthenogenesis of varying degrees of complexity. One of the most notable is that of the honeybee. Unfertilized eggs develop into drones, which are males. Fertilized eggs become worker females, which are kept in a nonreproductive state by secretions from the reproductive female, the queen bee.

Life cycles involving alternation between parthenogenesis and bisexual reproduction can be found in many species of Homoptera and Diptera (flies). Aphids (Homoptera) have a seasonal cycle consisting of a bisexual winter phase and a parthenogenetic summer phase; some species spend each phase on a different host plant. Temperature change, length of day, and food availability play major roles in initiating the phases. In the midge, a type of fly, the bisexual phase occurs in adults, and parthenogenesis takes place among the larvae (paedogenesis). Adult female midges deposit fertilized eggs, from which hatch larvae whose ovaries develop while the rest of the body retains a larval form. The ovaries of the larvae release eggs that enter the larval hemocoel (the space between body organs), where they undergo development while feeding on larval tissue. When sufficiently developed, the parthenogenetically produced young emerge either as larvae that continue parthenogenetic reproduction, forming larvae like themselves, or as male or female larvae that mature to become bisexually reproducing adults.

Provisions for the Developing Embryo

Invertebrates have developed a great many methods for protecting the fertilized egg and young embryo and for providing nutrients for the developing young. This is especially true of freshwater and terrestrial forms. Sponges and freshwater coelenterates, exposed to seasonal drying out, provide a tough covering for the eggs that prevents water loss. Many turbellarians envelop the eggs with a capsule and attach it to a hard surface, where it remains until the young emerge. Other turbellarians retain encapsulated eggs in the body until development is complete and the young emerge. All parasitic flatworms enclose their eggs in a protective capsule within which development occurs after it has left the parent's body. Most nematodes and rotifers do likewise, but a few species are ovoviviparous; i.e., the egg hatches in the mother's body. In many forms the amount of yolk provided in the egg and the nature of the egg capsule are correlated with annual seasons—summer eggs generally have less yolk and thinner capsules than do winter eggs. This is true also in a number of crustaceans. Freshwater and terrestrial annelids provide a cocoon for their young and often deposit it in a moist place. One group of leeches, however, does not form a cocoon; instead, the egg, surrounded by a protective membrane, is attached to the underside of the parent. As the young develop, the adult leech undulates its body so that water currents flow over the young. Presumably this serves as a means of aeration. Mollusks that live in freshwater may provide a protective covering for the eggs, or the eggs may be brooded by the female. Some pelecypods (bivalves) release mature eggs into their gill chambers; here the eggs are fertilized, and embryonic development is completed in a protected location. Cephalopods (e.g., squid, octopus) attach the eggs to a surface, and then continuously force jets of water over the egg masses, thereby keeping them free of debris and perhaps aerating them. Some echinoderms also brood the eggs until the young emerge.

Arthropods have a particularly wide range of methods for ensuring offspring survival. Brood pouches, common in branchiopods, isopods, and amphipods, are sometimes part of the carapace, or back plate. In other instances, expanded plates on the lower side (sternum) form the pouches. Crayfish cement the fertilized eggs to their swimmerets (modified appendages) and carry them about as they are brooded by the female. The most elaborate provisions for the embryo are found among terrestrial arthropods, especially insects. Although some species simply deposit their eggs and abandon them, many retain the encapsulated egg within the body during early development. Some are viviparous; that is, they bear living young. The eggs of certain species of scorpions have little or no yolk; the embryo is nourished by the parent in a manner similar to that in mammals—part of the scorpion oviduct becomes modified as a uterus for the embryo; another part lies close to the female's gut and absorbs nutritive substances that are conveyed to the developing young. A similar arrangement has evolved in some insects. Other viviparous insects nourish the larvae by glandular secretions from the uterine lining.

Reproductive Systems of Vertebrates

Gonads, Associated Structures and Products

The reproductive organs of vertebrates consist of gonads and associated ducts and glands. In addition, some vertebrates, including some of the more primitive fishes, have organs for sperm transfer or ovipository (egg-laying) organs. Gonads produce the gametes and hormones essential for reproduction. Associated ducts and glands store and transport the gametes and secrete necessary substances. In addition to these structures, most male and female vertebrates have a cloaca, a cavity that serves as a common terminal chamber for the digestive, urinary, and reproductive tracts and empties to the outside. In lampreys and most ray-finned fishes in which the cloaca is small or absent, the alimentary canal has a separate external opening, the anus. In some teleosts the alimentary, genital, and urinary tracts open independently. Hagfishes, which are closely related to the lampreys, have a short cloaca. In many vertebrates other than mammals, especially reptiles and birds, the cephalic, or head, end of the cloaca is partitioned by folds into a urinogenital chamber (urodeum) and an alimentary chamber (coprodeum) that open into a common terminal chamber (proctodeum). Above monotremes (e.g., platypus, echidna) the embryonic cloaca becomes completely partitioned into a urinogenital sinus conveying urine and the products of the gonads, and an alimentary pathway; the two open independently to the exterior.

Gonads arise as a pair of longitudinal thickenings of the coelomic epithelium and underlying mesenchyme (unspecialized tissue) on either side of the attachment of a supporting membrane, the dorsal mesentery, to the body wall. At first, gonadal ridges bulge into the coelom and are continuous with the embryonic kidney. The germinal epithelium covering the gonadal ridges gives rise to primary sex cords (medullary cords) that invade the underlying mesenchyme. These cords establish within the gonadal blastema (a tissue mass that gives rise to an organ) a potentially male component, the medulla. Secondary sex cords grow inward, spreading just beneath the germinal epithelium to form a cortex. If the gonad is to become a testis, only the medullary component differentiates. If the gonad is to become an ovary, only the cortex differentiates.

The length of an adult gonad depends, in part, upon the extent of gonadal-ridge differentiation. In cyclostomes (lampreys and hagfish), elasmobranchs (sharks, skates, and rays), and teleosts most of it differentiates, and the gonads extend nearly the length of the body trunk. In tetrapods (amphibians, reptiles, birds, and mammals), the cranial portion, at the anterior end, generally does not differentiate; in toads only the more caudal, or posterior, portion does so. The middle segment in toads of both sexes gives rise to a Bidder's organ containing immature eggs. In anurans (frogs and toads) and some lizards of both sexes, one segment of the gonadal ridge gives rise to yellow fat bodies that, especially in anurans, diminish in size just prior to the breeding season. In mammals, only the middle portion of the gonadal ridge differentiates.

Some vertebrate species have only one gonad, which may lie in the midline or on one side; the condition is more common among females. Adult cyclostomes of both sexes have one gonad. In lampreys it is in the middle of the body; in hagfishes it is on the right side. Birds are the only other major group of vertebrates in which most females have one gonad, the right ovary being typically absent. Male birds have a pair of testes, however. Exceptions to the condition of single ovaries among birds include members of the falcon family, in which more than 50 percent of mature hawks have two well-developed ovaries. In all bird species a small percentage of females probably have two ovaries; reported instances include owls, parrots, sparrows, and doves, with estimates for doves ranging from 5 percent to 25 percent. A few teleosts and viviparous elasmobranchs have only one ovary; in sharks the right one is usually presents, in rays, the left. In amniotes (i.e., reptiles, birds, and mammals) unpaired gonads are unusual. Some lizards have one testis, and some female crocodiles have one ovary. Among mammals, the platypus usually has only a left ovary, and some bat species (family Vespertilionidae) have only the right.

One of two explanations may account for unpaired gonads: the paired embryonic gonadal ridges may fuse to form a median gonad—as in lampreys and the perch—or only one gonadal ridge may receive immigrating primordial germ cells (immature sperm or eggs), with the result that the opposite gonad does not develop—as in chickens and ducks. Both gonadal ridges have been reported to exhibit an equal number of primordial germ cells in embryonic hawks, and these typically have two ovaries.

Among lower vertebrates, mature gonads sometimes produce both sperm and eggs. Hermaphroditism is more general in cyclostomes and teleosts than in other fishes. A teleost may function as a male during the early part of its sexual life and as a female later. In some teleost families sperm and eggs mature simultaneously but in different regions of the gonad. These fish normally function as males during one season and as females the next. Cyclostomes generally are ambisexual during juvenile life—i.e., immature male and female sex cells exist side by side, or, as in Myxine, the anterior part of the immature gonad may be ovary and the caudal part, the testis. It is thought that cyclostomes normally become unisexual at maturity. Hermaphroditism is uncommon among amphibians, although it frequently occurs as an anomaly. In vertebrates above amphibians, true hermaphroditism probably does not exist.

Both male and female duct systems are occasionally absent. In cyclostomes, a few elasmobranchs, and some teleosts, such as salmon, trout, and eels, the gametes are propelled toward the posterior within the coelom, often by cilia (minute hairlike structures), and exit via a pair of funnel-like genital pores near the base of the tail. In cyclostomes, the pores lead to a sinus, or cavity, within a median papilla (i.e., a fingerlike structure) and are open only during breeding seasons.

Male Systems

Testes

In anurans, amniotes (reptiles, birds, and mammals), and even some teleosts, testes are composed largely of seminiferous tubules—coiled tubes, the walls of which contain cells that produce sperm—and are surrounded by a capsule, the tunica albuginea. Seminiferous tubules may constitute up to 90 percent of the testis. The tubule walls consist of a multilayered germinal epithelium containing spermatogenic cells and Sertoli cells, nutritive cells that have the heads of maturing sperm embedded in them. Seminiferous tubules may begin blindly at the tunic, or outermost tissue layer, and pass toward the centre, becoming tortuous before emptying into a system of collecting tubules, the rete testis. Such an arrangement is characteristic of frogs. In certain amniotes—the rat, for example—the tubules may be open ended, running a zigzag course from the rete to the periphery and back again. The average length of such tubules is 30 centimetres (12 inches), and they seldom communicate with each other. In many mammals the tubules are grouped into lobules separated by connective-tissue septa, or walls. The arrangement permits the packing of an extensive amount of germinal epithelium into a small space. In immature males and in adult males between breeding seasons, the tubules are inconspicuous and the epithelium is inactive; in some species, however, spermatogenesis, or production of sperm, proceeds at a variable pace throughout the year. An active epithelium may exhibit all stages of developing sperm. The lumen, or tubule cavity, contains the tails of many sperm (the heads of which are embedded in Sertoli cells), free sperm, and fluid that is probably resorbed. In mammals, in any single zone along a tubule, all sperm are at the same stage of maturation; adjacent zones contain different generations of sperm, and a period of sperm formation and discharge is followed by an interval of inactivity.

In cyclostomes, most fishes, and tailed amphibians the germinal epithelium is arranged differently. Instead of seminiferous tubules there are large numbers of spermatogonial cysts (also called spermatocysts, sperm follicles, ampullae, crypts, sacs, acini, and capsules) in which sperm develop, but in which the epithelium is not germinal. Spermatogenic cells migrate into the cysts from a permanent germinal layer, which, depending on the species, may lie among cysts at the periphery of the testes or in a ridge along one margin of the testis. After invading the thin nongerminal epithelium of a cyst, spermatogenic cells multiply, producing enormous numbers of sperm. The cysts become greatly swollen and whitish in colour; the entire testis also swells and has a granular appearance. As sperm mature, they separate from the epithelium and move freely in the cystic fluid. Finally, the cysts burst, and the sperm are shed into ducts. In the case of cyclostomes and a few teleosts the sperm are shed into the coelom. The cysts, totally emptied, collapse. Then either they are replaced by new ones, or they become repopulated by additional spermatogenic cells. It is not yet known which of these processes occurs.

Testicular stroma, which fills the spaces between seminiferous tubules or spermatogenic cysts, consists chiefly of connective tissue, blood and lymphatic vessels, and nerves; it is more abundant in some vertebrates than in others. Glandular Leydig (interstitial) cells are also present in most, if not all, vertebrates. Thought to be a primary source of androgens, or male hormones, Leydig cells are not always readily distinguishable, and, in some bird species, they may be seen only with the electron microscope. The capillary system of the rat testis, and probably that of many other vertebrates, is such that blood that has bathed the Leydig cells flows to the tubules; it is thus probable that Leydig cell hormones have an immediate effect on the germinal epithelium.

Testes in vertebrates below mammals lie within the body. This is also true of many, sometimes all, members of the mammalian orders Monotremata, Insectivora, Hyracoidea, Edentata, Sirenia, Cetacea, and Proboscidea. Some male mammals—most marsupials, ungulates, carnivores, and primates after infancy—have a special pouch (scrotum) that the testes occupy permanently. A few mammals have a pouch into which the testes descend and from which they can be retracted by muscular action. These include a few rodents such as ground squirrels; most, if not all, bats; and some primitive primates (loris, potto). The scrotum consists of two scrotal sacs, each connected to the abdominal cavity by an inguinal canal lined with the peritoneal membrane. The canals are the path of descent (and retraction) of the testes to the sacs. In descending, the testes carry along a spermatic duct, blood and lymphatic vessels, and a nerve supply wrapped in peritoneum and constituting, collectively, the spermatic cord. Rabbits, most rodents, and some insectivores, which lack scrotal sacs, have instead a wide inguinal canal into which the testes may be drawn and from which they are retracted when in danger of injury. In these mammals, descended testes cause a temporary bulge in the perineal region (i.e., between the anus and the urinogenital opening). In a small number of mammals, the testes permanently occupy the perineal location.

The scrotum is a temperature-regulating device. Warm blood approaching the testis comes close to the vessels carrying cool blood leaving the testis, so that the blood approaching the testis is cooled; the vessels form an intricate vascular network (pampiniform plexus) within the spermatic cord. Failure of both testes to enter the scrotal sacs (cryptorchidism) results in permanent sterility. In cold weather two sets of muscles, the dartos and cremasteric, pull the testes close to the body. The dartos lies between the two scrotal sacs and is attached to the scrotal skin. The cremaster, wrapped around the spermatic cord, is an extension of the abdominal wall musculature. It retracts the testis. Birds, like mammals, are homoiothermic (warm-blooded), and their testes are near air sacs (extensions of incurrent respiratory tubes). Air in the sacs may help regulate the temperature of the testes.

Ducts

The male duct system begins as the rete testis, a network within the testis of thin-walled ductules, or minute ducts that collects sperm from the seminiferous tubules. The rete is drained by a number of small ducts—usually fewer than ten—called the vasa efferentia, which are modified kidney tubules. In some fishes and amphibians the vasa efferentia connect the testes with the cranial (anterior) end of the kidneys. In anamniotes (e.g., fish and amphibians), therefore, except teleosts, the ducts that drain the kidneys usually drain the testes also. In most amphibians these ducts pass caudad, or posteriorly, to empty independently into the cloaca; in some fishes they pass through a median urinogenital papilla.

Although drainage of the testis and the kidney by the same duct is a basic pattern, there has been a tendency in many vertebrates toward separate spermatic and urinary ducts. This tendency is manifested in one of two ways among anamniotes. In many sharks and in some amphibians (Plethodontidae, Salamandridae, Ambystomatidae), the embryonic kidney duct ultimately drains the testis, and one or more new ducts (ureters) drain the adult kidney. On the other hand, in the primitive fish Polypterus and in most teleosts, the embryonic kidney duct drains the adult kidney, and a new duct arises to drain the testis. Many degrees of separation of the two ducts occur in anamniotes, from the condition of the sturgeon, in which the spermatic duct unites with the urinary duct far toward the head, to the condition in Esox (a pike), in which spermatic and urinary ducts empty independently to the exterior.

In amniotes, the mesonephric kidney is a temporary structure confined to the embryo, but the mesonephric duct persists in the adult male as a sperm duct. A separate ureter drains the adult kidney. The spermatic and urinary ducts empty independently into the cloaca except in mammals above monotremes, in which they are confluent with the urethra. The epididymis of amniotes, a highly tortuous duct draining the vasa efferentia, usually serves as a temporary storage place for sperm; it is small in birds and large in turtles. In mammals, the first part of the epididymis consists of a head, body, and tail that wrap around the testis; it gradually straightens to become the spermatic duct. The epididymis secretes substances that prolong the life of stored sperm and increase their capacity for motility.

In all vertebrates certain regions of the spermatic duct are lined by cilia and a variety of secretory epithelial cells. One end may enlarge to form a sperm reservoir or secrete seminal fluid. In the catfish Trachycorystes mirabilis secretions of the spermatic duct form a gelatinous plug in the female similar to the vaginal plug of mammals. A seminal glomulus in birds functions as a sperm reservoir. In some mammals an enlargement of the spermatic duct called the ampulla contributes to the seminal fluid and stores sperm. Small mucous glands (of Littré) and other glandular structures open into the urethra along its length. Cloacal glands in basking sharks and many salamanders form a jelly that encloses sperm in a spermatophore. Cloacal glands of some lizards produce secretions called pheromones. The siphon sac of elasmobranchs is one of the few accessory sex glands that are a separate organ in animals below mammals. It extends as an elongated pocket into the pelvic fin and secretes nutritive mucus that enters the female reproductive tract with sperm.

Accessory Glands

Accessory sex glands that are conspicuous outgrowths of the genital tract are almost uniquely mammalian. The major mammalian sex glands include the prostate, the bulbourethral, and the ampullary glands, and the seminal vesicles. All are outgrowths of the spermatic duct or of the urethra and all four occur in elephants and horses and in most moles, bats, rodents, rabbits, cattle, and primates. A few members of these groups lack ampullary glands, or ampullary glands and seminal vesicles. Cetaceans (whales, porpoises) have only the prostate, as do some carnivores, including dogs, weasels, ferrets, and bears.

The prostate, the most widely distributed mammalian accessory sex gland, is absent only in Echidna (a marsupial) and a few carnivores. It empties into the urethra by multiple ducts. Many rodents, insectivores, and lagomorphs have three separate prostatic lobes; in a few mammals (some primates and carnivores) the prostate is a single mass with lobules and encircles the urethra at the base of the bladder. In a few mammals (e.g., opossum), the prostate is not a compact mass but a partly diffuse gland. In many rodents (e.g., rat, guinea pig, mouse, and hamster) and some other mammals, the semen coagulates quickly after ejaculation as a result of a secretion from a male coagulating gland, which is usually considered part of the prostatic mass. Coagulated semen forms a vaginal plug that temporarily prevents copulation.

Bulbourethral (Cowper's) glands arise from the urethra near the penis and are surrounded by the muscle of the urethra or penis. Typically, there is one pair, but as many as three (marsupials) may be found. The glands, small in man, large in rodents, elephants, and some ungulates including pigs, camels, and horses, are absent in cetaceans, mustelids (e.g., mink, weasel), sirenians (manatees, dugongs), pholidotans (pangolins), some edentates, and carnivores such as walrus, sea lion, bear, and dog.

Although many mammals have an ampullary swelling on the spermatic duct near the urethra, only a small number form a separate ampullary gland as an outgrowth of the duct. It is very large in some bats, absent in many mammalian orders, and variable in the rest. Although common in rodents, it is absent in guinea pigs and some strains of mice.

Seminal vesicles are paired, typically elongated and coiled fibromuscular sacs that empty into either the spermatic duct or the urethra. Absent in monotremes, marsupials, carnivores, cetaceans, and in some insectivores, chiropterans, and primates, seminal vesicles are exceptionally large in rhesus monkeys and small in man. They are absent in domesticated rabbits, small or rudimentary in cottontails, large in armadillos, and variable in sloths. They contribute the sugar fructose and citric acid to the semen but do not serve as sperm reservoirs.

Female Systems

Ovaries

Ovaries lie within the body cavity and are suspended by a dorsal mesentery (mesovarium), through which pass blood and lymph vessels and nerves. Primitive vertebrate ovaries occur in the hagfish, in which a mesentery-like fold of gonadal tissue stretches nearly the length of the body cavity. Unique in the hagfish is the fact that functional ovarian tissue occupies only the forward half of the gonadal mass, the rear part containing rudimentary testicular tissue. In most fishes except very primitive forms, the ovaries are similarly elongated. In tetrapods other than mammals, the ovaries are usually confined to the middle third or half of the body cavity, particularly during nonbreeding seasons. The ovaries of mammals undergo moderate caudal displacement, finally coming to lie between the kidney and the pelvis.

The appearance of an ovary depends on many factors—e.g., whether one egg or thousands are discharged (ovulated); whether the eggs are immature or ripe; whether mature eggs are small or large; or whether pigments occur in the egg cytoplasm, such as those responsible for yellow yolk. Other factors also affect the appearance of the ovary: the season of the year in seasonal breeders (the ovary enlarges during breeding seasons, diminishes in size between seasons); the age of the animal (whether juvenile, reproductively active, or senile, particularly in birds and mammals); and the fate of ovulated, or discharged, egg follicles, or sacs.

The ovaries are covered with a germinal epithelium that is continuous with the peritoneum lining the body cavity. The term germinal epithelium is inappropriate because in most adults it contains no germ cells, these having moved deeper into the ovary. In hagfishes and amphibians, cells that give rise to eggs are known to occur in the germinal epithelium, and it may be that the germinal epithelium in a few other vertebrates contains similar cells. The germinal epithelium undergoes cell division, however. This is particularly true of species in which enormous expansion of the ovary occurs each breeding season. Beneath the epithelium is a layer of connective tissue, the tunica albuginea, which is much thinner than that surrounding the testes.

A typical vertebrate ovary consists of cortex and medulla. The cortex, immediately internal to the tunica albuginea, contains future eggs and, at one time or another, eggs in ovarian follicles (i.e., developing eggs); it undergoes fluctuations in size and appearance that correlate with stages of the reproductive cycle. The cortex also contains remnants of ovulated follicles and, in mammals, clusters of interstitial cells that, in some species, are glandular. The cortical components are embedded in a supportive framework of connective, vascular, and neural tissue constituting the stroma. Internal to the cortex is the medulla, consisting of blood and lymph vessels, nerves, and connective tissue. The medulla, which contains no germinal elements, exhibits no significant cyclical activity, is usually inconspicuous, is continuous with the dorsal mesentery and, in cyclostomes, is hardly distinguishable from the latter. The mammalian medulla, on the contrary, is almost completely surrounded by cortex and converges on the mesovarium (i.e., the part of the peritoneum that supports the ovary) at a narrow hilus, at which nerves and vessels enter the ovary. In the medulla of the mammalian ovary near the hilus are small masses of blind tubules or solid cords—the rete ovarii—which are homologous (i.e., of the same embryonic origin) with the rete testis in the male. The microscopic right ovary of birds usually consists only of medullary tissue.

Ovaries are characterized as saccular, hollow, lacunate (i.e., compartmented), or compact. The ovary of many teleosts, especially viviparous ones, contains a permanent cavity, which is formed during ovarian development when an invagination of the ovarian surface traps a portion of the coelom. The cavity is therefore unique in that it is lined by germinal epithelium. The lining develops numerous ovigerous folds that project into the lumen and greatly increase the surface area for proliferation of eggs. In most other teleosts, a temporary ovarian cavity develops after each ovulation, when the shrinking cortex withdraws from the outside ovarian wall along one side of the ovary. The resulting cavity is obliterated as eggs of the next generation enlarge. The permanent and temporary cavities of teleost ovaries and a similar cavity in garfish ovaries are continuous with the lumen of the oviduct, and eggs are shed into them. The ovaries of other fishes lack cavities and are characterized as compact. The amphibian ovary, which contains six or more central, hollow sacs that give it a lobed appearance, is characterized as saccular. The sacs are formed when the embryonic medullary and rete cords become hollow and coalesce. Maturing eggs bulge into the sacs but are not shed into them. The ovaries of reptiles, birds, and monotremes have cavities homologous to those in amphibians; the number of medullary spaces in the adults is considerably larger, however, so that the ovaries contain an extensive network of fluid-filled cavities (lacunae). Such ovaries are characterized as lacunate. The ovaries of mammals above monotremes are compact, having no medullary cavities.

An ovarian follicle consists of an oocyte, or immature egg, surrounded by an epithelium, the cells of which are referred to variously as follicular, nurse, or granulosa cells. In cyclostomes, teleosts, and amphibians, the epithelium is one layer thick. In the hagfish and those vertebrates in which the oocyte receives heavy deposits of yolk (elasmobranchs, reptiles, birds, and monotremes), the epithelium appears to be two cells thick, apparently the result of layering of nuclei in a simple columnar epithelium (i.e., epithelium consisting of relatively "tall" cells). Above monotremes the follicular epithelium appears to be many cells thick; in at least one species, however, this is considered an artifact, and all granulosa cells are said to extend between the outer boundary of the epithelium and the oocyte.

The follicular epithelium originates as a few flattened cells derived from the germinal epithelium. Primary follicles are usually situated just under the tunica albuginea; secondary follicles lie deeper in the cortex. The primitive role of the follicular cells appears to be the secretion of the yolk-forming material onto or into the oocyte. Evidence from mammals indicates that the follicular cells may also have a role in converting substances produced elsewhere into female hormones, or estrogens. In some hibernating bats the granulosa cells are filled with glycogen, or animal starch, which may be a source of energy. Mammalian follicles above monotremes are unique in that they develop a fluid-filled cavity (antrum) within the granulosa layer. During antrum formation cell division of the granulosa cells increases and fluid-filled spaces develop among the cells. The spaces coalesce to form the antrum. Under the influence of pituitary gonadotropic hormones, many antral follicles thereafter continue to grow, forming large so-called Graafian follicles—less than 400 microns, or 0.4 millimetre (0.16 inch), in diameter in large mammals, 150–200 microns, or 0.15–0.2 millimetre (0.006–0.008 inch), in small ones. Graafian follicles contain mature eggs and appear as large blisters on the ovary. At this stage the ovum, suspended within the fluid of the antrum (liquor folliculi) by a slender stalk of granulosa cells, is surrounded by a cluster of these cells, the cumulus oophorus, or discus proligerus. The remaining follicular cells form a thin wall surrounding the antrum. Antra are lacking in a few insectivores (Hemicentetes, Euriculus) because the granulosa cells swell and multiply to form corpora lutea, masses of yellow tissue. In the bat Myotis the antrum is likewise compressed and disappears just before discharge of the egg, or ovulation.

In all vertebrates, oocytes that have begun to grow and mature may, at any time until just before ovulation, cease development and undergo atresia, or degeneration. This is a normal process that reduces the number of eggs ovulated. In small laboratory rodents, atresia takes place in 50 percent of the Graafian follicles in each ovary one or two days before ovulation, thus reducing the number of ovulatable eggs by 50 percent. A similar reduction takes place in hagfish prior to ovulation. Atretic follicles eventually become lost in the stroma of the cortex of the ovary. In mammals especially, follicles lacking oocytes and antra, called anovular follicles, as well as polyovular follicles (i.e., containing more than one oocyte), occasionally occur.

The ovarian follicle of vertebrates, commencing with hagfish, is surrounded by a theca, or sheath, composed of two concentric layers of stromal cells. The outer layer (theca externa) is chiefly connective tissue but may contain smooth muscle fibres. The inner layer (theca interna) has more blood vessels and, in vertebrates that produce heavily yolked eggs, the largest vessels carry venous blood. In these species the cell membranes of the oocyte and granulosa cells have many microvilli (i.e., fingerlike projections), which probably facilitate transport of substances important in yolk formation from the blood vessels to the egg. Mature follicles in the marsupial Dasyatus are said to lack theca, and in some bats only one thecal layer has been described.

During the growth phase, eggs in species with massive amounts of yolk may increase in size 106 (1,000,000) or more times as a result of vitellogenesis (deposit of yolk). In goldfish, on the other hand, when vitellogenesis commences, the egg has a diameter of 150 microns (0.15 millimetre [0.006 inch]); that of the mature egg is only 500 microns (0.5 millimetre [0.02 inch]). Mammalian eggs contain little yolk and very little in size. Oogonia (i.e., cells that form oocytes) of the golden hamster average 15 microns (0.015 millimetre [0.0006 inch]) in diameter, and eggs in Graafian follicles average 70 microns (0.07 millimetre [0.003 inch]). The mature eggs of horses and humans are approximately the same size—somewhat less than 150 microns. In seasonally breeding oviparous fishes and amphibians, all eggs are usually in the same stage of development, and the ovary grows to a mature state quite rapidly as a result of growth of the eggs, which frequently number more than 1,000,000. Such ovaries distend the body wall when mature; following spawning, the ovaries shrink rapidly to inconspicuous bodies consisting mainly of oogonia, immature oocytes, and a few stromal cells. In reptiles and birds, ovarian weight also is high in proportion to body weight during egg-laying seasons. The weight of the ovary of the starling, for example, may increase from eight milligrams in early winter to 1,400 milligrams immediately before ovulation. The mature eggs of reptiles and birds are unique in that they are suspended from the ovary by a short stalk (pedicle). The stalk contains a cortex with additional oocytes in various stages of development and extensions of vessels and nerves. Full growth of the follicle in reptiles and birds requires only a few days or weeks (nine days in the domestic hen). In mammals, the ratio of ovarian weight to body weight varies insignificantly throughout the reproductive life of the female, and follicles in many stages of development are constantly present.

Vertebrate eggs are almost universally shed into the coelom or into a subdivision thereof, from which they enter the female reproductive tract. Even in those teleosts in which the eggs are shed into an ovarian cavity, the latter is often of coelomic origin. In many mammals a membranous sac of peritoneum, the ovarian bursa, traps part of the coelom in a chamber along with the ovary. The bursal cavity (periovarian space) may be broadly open to the main coelom, completely closed off from the coelom, or in communication with the coelom by a narrow, slitlike passage. The bursa, moderately developed in lower primates and catarrhines (Old World monkeys), is poorly developed in man. In horses, one edge of the ovary contains a long groove (ovulation fossa) into which all eggs is shed; the groove is found in a cleftlike ovarian bursa. The ovarian bursa increases the probability that all ovulated eggs will enter the oviduct.

The process of ovulation has been described for all vertebrate classes. Elasmobranchs, reptiles, and birds have massively yolked eggs. As ovulation approaches, the fimbria (i.e., frills, or fringes) of the membranous and muscular funnel surrounding the entrance to the oviduct wave in a gentle, undulating motion. An egg that is nearly free of the ovary is grasped and partially encompassed by the fimbria; when the egg is freed, the fimbria draws the egg into the funnel. At this time, the egg has little shape and is partly squirted and partly flows into the oviduct; never completely free in the coelom, its chances of not entering the oviduct are small. In the case of moderately or poorly yolked eggs cilia help to sweep the eggs into the ostium, or opening, of the oviduct. During ovulation in Japanese rice fish, Oryzias latipes, a tiny papilla, or fingerlike process, develops on the surface of a bulging mature follicle in the centre (stigma) of the follicle. The follicle becomes thin at the stigma, an aperture appears, and the egg rolls out. In rabbits this process differs only in detail. During the final 20 minutes before ovulation in rabbits, some of the tiny blood vessels surrounding the stigma rupture, and a small pool of blood forms under the apex of the cone-shaped papilla. The follicular wall shortly gives way at the apex, and follicular fluid oozes from the opening, followed soon after by the egg. The ovulated mammalian egg typically is surrounded by a layer of columnar follicular cells, the corona radiata; but it is naked in some insectivores and some marsupials. Following ovulation in all vertebrates, the ovary may become smaller, become modified for maintenance of pregnancy, or proceed to form additional eggs.

The process of ovulation in vertebrates has been documented, but the immediate causes remain to be clarified. It is almost certain that an ovulatory hormone is secreted by the pituitary gland (i.e., the so-called master endocrine gland) of all vertebrates. It is highly probable that breakdown of very small fibres that bind the follicular cells together may occur at the stigma, weakening the follicular wall at that location. Hormones from the ovary and other sources may play a role, as May neurohormones, which are hormones released at nerve endings. Rhythmic contractions of the entire ovary occur at ovulation in many vertebrates and have been described in rabbits. The role of mechanical pressure within the follicle, however, is not understood. Ovulation in most mammals (spontaneous ovulators) occurs cyclically as a result of the spontaneous release of the ovulatory hormone. In a few mammals (reflex ovulators) the stimulus of copulation is essential for release of the ovulatory hormone.

Striking postovulatory changes take place in the follicles of mammals and, to lesser degrees, of lower vertebrates. Blood vessels from the theca interna invade the ovulated follicles; the granulosa cells divide, enlarge, accumulate fats, and obliterate any remnants of the collapsed antra. Thereafter, they are known as lutein cells. Theca interna cells undergo changes identical to those of the granulosa cells. The result in mammals is the formation of solid masses called corpora lutea, recognizable as prominent reddish-yellow bulges on the ovary. Corpora lutea produce the hormone progesterone, which is essential for the maintenance of pregnancy. The conversion of postovulatory follicles into structures more or less resembling mammalian corpora lutea has been demonstrated in numerous viviparous reptiles, amphibians, and elasmobranchs; in certain other fishes, including cyclostomes; and in some oviparous amphibians and reptiles. In birds, the postovulatory follicle shrinks, and identifiable corpora lutea do not develop, although some granulosa cells accumulate lipids of unknown significance.

Tracts

The female reproductive tract consists of a pair of tubes (gonoducts) extending from anterior, funnel-like openings (ostia) to the cloaca, except as noted below. The gonoducts are specialized along their length for secretion of substances added to the eggs; for transport, storage, nutrition, and expulsion of eggs or the products of conception; and, in species with internal fertilization, for receipt, transport, storage, and nutrition of inseminated sperm. The predominately muscular tracts are lined by a secretory epithelium and ciliated over at least part of their length. Fusion of the caudal (tail) ends of the paired ducts may occur. Gonoducts are absent in cyclostomes and a few gnathostome fishes that have abdominal pores. A few vertebrates have only one functional gonoduct.

Gonoducts in lungfishes and amphibians are coiled muscular tubes that are ciliated over most of their length. Only occasionally do they unite caudally in a genital papilla before opening into the cloaca. During breeding seasons their diameter increases severalfold because of the highly active secretory epithelium. Between breeding seasons they are small. In some anurans (frogs, toads), such as Rana, the lower end of each gonoduct is expanded to form an ovisac, in which ovulated eggs are stored until spawning; the tube between the ostium (funnel-like opening) and ovisac is the oviduct. In viviparous amphibians the young develop in the ovisac. In amphibians, numerous multicellular glands extend deep into the lining of the female tract. Six successive glandular zones have been described in some urodeles, and these secrete six different gelatinous substances upon the egg. Female urodeles often have convoluted tubular outpocketings of the cloaca called spermatheca; they temporarily store sperm liberated from the male spermatophore.

The two gonoducts of elasmobranchs share a single ostium, a trait found only in Chondrichthyes. The ostium is a wide caudally directed funnel supported in the falciform ligament, which is attached to the liver. The role of the fimbria of the ostium at ovulation has been described (see above Ovaries). Two oviducts pass forward from the ostium to the septum transversum (i.e., between the heart and abdominal cavities), curve around one end of the liver, then pass posteriorly on each side. Approximately midway between ostium and uterus each oviduct has a shell (nidamental) gland. Fertilization takes place above the shell gland, which may be immense or almost undifferentiated. Half of the shell gland secretes a substance high in protein content (albumen), and the other half secretes the shell—delicate in viviparous forms, thick and horny in most oviparous species. Horny shells may have spiral ridges and many long tendrils, which entwine about an appropriate surface after the egg is deposited. In the viviparous shark Squalus acanthias several eggs pass one after the other through the shell gland, where they are enclosed in one long delicate membranous shell that soon disintegrates. Beyond the shell gland the oviducts terminate in an enlargement, which, in viviparous species, serves as a uterus. An oviducal valve may be found at the junction of oviduct and uterus. Although the two uteri usually open independently into the cloaca, they occasionally unite to form a bicornuate (two-horned) structure. In immature females, the uterus may be separated from the cloaca by a hymen, or membrane. The tract enlarges enormously during the first pregnancy and does not thereafter fully regress to its original size.

The gonoducts of lower ray-finned fishes resemble those of lungfish, but those of gars and teleosts are exceptional in that the oviducts are usually continuous with the ovarian cavities. A median genital papilla receives the oviducts in teleosts, and the papilla is sometimes elongated to form an ovipositor. European bitterlings deposit their eggs in a mussel by means of the ovipositor, and female pipefish and sea horses deposit them in the brood pouch of a male.

With certain modifications, the gonoducts of reptiles and birds are comparable to those of lower vertebrates. Crocodilians, some lizards, and nearly all birds have one gonoduct; the other is not well developed. Even in birds of prey having two functional ovaries, the right oviduct is sometimes undeveloped. The tracts of reptiles generally show less regional differentiation than do those of birds. The oviduct funnel (ostium) in birds forms the chalazae—two coiled, springlike cords extending from the yolk to the ends of the egg. In both reptiles and birds, much of the length of the female tract is oviduct. This region, called the magnum in birds, secretes albumen; lizards and snakes do not form albumen. Behind the albumen-secreting region is a shell gland. In lizards, the gland is midway along the tract. In birds, the shell gland is at the posterior end, has thick muscular walls, and is often inappropriately called a uterus. It is preceded by a narrow region, or isthmus, which secretes the noncalcareous, or soft, membranes of the shell. The shell gland leads to a narrow muscular vagina that empties into the cloaca. The vagina secretes mucus that seals the pores of the shell before the egg is expelled. Special vaginal tubules (spermatheca) store sperm over winter in some snakes and lizards; seminal receptacles have been described in the oviduct funnel in some snakes. In birds, sperm storage glands (sperm nests) often occur in the funnel and at the uterovaginal junction. In lizards and birds, ovulation does not usually occur into a tract already containing an egg. Some lizards shed very few eggs per season; the gecko, for example, sheds only two.

The female reproductive tracts of monotremes, the egg-laying mammals, consist of two oviducts, the lower ends of which are shell glands. These open into a urinogenital sinus, which, in turn, empties into a cloaca. Marsupials have two oviducts, two uteri (duplex uterus), and two vaginas. The upper parts of the vaginas unite to form a median vagina that may or may not be paired internally. Beyond the median vagina, the vaginas are again paired (lateral vaginas) and lead to a urinogenital sinus. The posterior end of the pouchlike median vagina is separated from the forward end of the urinogenital sinus by a partition. When the female is delivering young, the fetuses are usually forced through the partition and into the urinogenital sinus, bypassing the lateral vaginas. The ruptured partition may remain open thereafter, resulting in a pseudovagina. It closes in opossums, and in kangaroos both the median and lateral routes may serve as birth canals. The lateral vaginas in marsupials receive the forked tips of the male penis. Fertilization in all mammals takes place in the oviducts (Fallopian tubes).

In eutherian mammals (i.e., all mammals except monotremes and marsupials), with exceptions noted below, female reproductive tracts beyond the ostia (oviduct funnels) consist of two narrow and somewhat tortuous Fallopian tubes, two large uterine horns (each of which receives a Fallopian tube), a uterine body, and one vagina. Fallopian tubes often have a short dilated ampulla, or saclike swelling, just beyond the ostium. Implantation of the egg occurs only in the uterine horns; the embryos become spaced equidistant from one another in both horns even if only one ovary has ovulated. In some species one horn is rudimentary—the left in the impala (an African antelope)—and the embryos become implanted in the other horn, even though both ovaries ovulate. The body of the uterus in some mammals (e.g., rabbits, elephants, aardvarks; some rodents, bats, insectivores) contains two separate canals (bipartite uterus). In other mammals (ungulates, many cetaceans, most carnivores and bats) the body of the uterus has one chamber into which the two horns empty (bicornuate uterus). There are numerous intermediate conditions between the bipartite and bicornuate condition. Apes, monkeys, and man have no horns, and the Fallopian tubes empty directly into the body of the uterus (simplex uterus). In all mammals, the uterine body tapers to a narrow neck (cervix). The opening (os uteri) into the vagina is guarded by fleshy folds (lips of the cervix). The vagina in eutherian mammals other than rodents and primates terminates in a urinogenital sinus that opens to the exterior by a urinogenital aperture. In some rodents and in higher primates the vagina opens directly to the exterior. In the young of many species a membrane, the hymen closes the vaginal opening. In guinea pigs the hymen reseals the opening after each reproductive period. Sperm are stored over winter in the uterus of some bats and in vaginal pouches in others.

Accessory Glands

Female mammals have fewer accessory sex glands than males, the most prominent being Bartholin's glands and prostates. Bartholin's (bulbovestibular) glands are homologues of the bulbourethral glands of males. One pair usually opens into the urinogenital sinus or, in primates, into a shallow vestibule at the opening of the vagina. Prostates develop as buds from the urethra in many female embryos but often remain partially developed. They become well developed, however, in some insectivores, chiropterans, rodents, and lagomorphs, although their function is obscure. A variety of glands (labial, preputial, urethral) are found in the mucosa, or mucous membrane. Glands in the uterine mucosa provide nourishment for embryos before implantation. Cervical uterine glands secrete mucus that lubricates the vagina, which has no glands.

References

- Muscular-system, body-systems, physiology: poultryhub.org, Retrieved 24, March 2020
- Muscle-in-soft-animals, muscle, science: britannica.com, Retrieved 27, Feb 2020
- Skeletal-system: veteriankey.com, Retrieved 18, July 2020
- Digestion-unit-23271034: mishicotffa.org, Retrieved 02, April 2020
- The-main-excretory-system-in-animals-is-the-urinary-system, guiada-o-a-excretor, organos: mmegias.webs.uvigo.es, Retrieved 12, August 2020
- Animal-reproductive-system, science: britannica.com, Retrieved 30, August 2020

Chapter 4

Nervous Systems and Important Sensory Organs

The nervous system integrates and controls the various functions of the body, and detects and reacts to stimuli with the help of the sensory organs. The nervous system coordinates the actions and sensory information by transmitting signals to and from different parts of the body. The topics elaborated in this chapter will help in gaining a better perspective of the sensory organs and the nervous system in animals.

Lateral Line

This is an example of lateral line of a fish.

The lateral line is a system of sense organs found in aquatic vertebrates, used to detect movement, vibration, and pressure gradients in the surrounding water. The sensory ability is achieved via modified epithelial cells, known as hair cells, which respond to displacement caused by motion and transduce these signals into electrical impulses via excitatory synapses. Lateral lines serve an important role in schooling behavior, predation, and orientation. For example, fish can use their lateral line system to follow the vortices produced by fleeing prey. Lateral lines are usually visible as faint lines of pores running lengthwise down each side, from the vicinity of the gill covers to the base of the tail. In some species, the receptive organs of the lateral line have been modified to function as electroreceptors, which are organs used to detect electrical impulses, and as such, these systems remain closely linked. Most amphibian larvae and some fully aquatic adult amphibians possess mechanosensitive systems comparable to the lateral line.

Function

The lateral line system allows the detection of movement, vibration, and pressure gradients in the water surrounding an animal, providing spatial awareness and the ability to navigate in the environment. This plays an essential role in orientation, predatory behavior, defense, and social schooling.

The small holes on the head of this Northern pike (*Esox lucius*) contain neuromasts of the lateral line system.

The lateral line system is necessary to detect vibrations made by prey, and to orient towards the source to begin predatory action. Fish are able to detect movement, produced either by prey or a vibrating metal sphere, and orient themselves toward the source before proceeding to make a predatory strike at it. This behavior persists even in blinded fish, but is greatly diminished when lateral line function was inhibited by $CoCl_2$ application. Cobalt chloride treatment results in the release of cobalt ions, disrupting ionic transport and preventing signal transduction in the lateral lines. These behaviors are dependent specifically on mechanoreceptors located within the canals of the lateral line.

The role mechanoreception plays in schooling behavior was demonstrated in a 1976 study. A school of *Pollachius virens* was established in a tank and individual fish were removed and subjected to different procedures before their ability to rejoin the school was observed. Fish that were experimentally blinded were able to reintegrate into the school, while fish with severed lateral lines were unable to reintegrate themselves. Therefore, reliance on functional mechanoreception, not vision, is essential for schooling behavior. A study in 2014 suggests that the lateral line system plays an important role in the behavior of Mexican blind cave fish (Astyanax mexicanus).

Anatomy

The major unit of functionality of the lateral line is the neuromast. The neuromast is a mechanoreceptive organ which allows the sensing of mechanical changes in water. There are two main varieties of neuromasts located in animals, canal neuromasts and superficial or freestanding neuromasts. Superficial neuromasts are located externally on the surface of the body, while canal neuromasts are located along the lateral lines in subdermal, fluid filled canals. Each neuromast consists of receptive hair cells whose tips are covered by a flexible and jellylike cupula. Hair cells typically possess both glutamatergic afferent connections and cholinergic efferent connections. The receptive hair cells are modified epithelial cells and typically possess bundles of 40-50 microvilli "hairs" which function as the mechanoreceptors. These bundles are organized in rough "staircases" of hairs of increasing length order. This use of mechanosensitive hairs is homologous to the functioning of hair cells in the auditory and vestibular systems, indicating a close link between these systems.

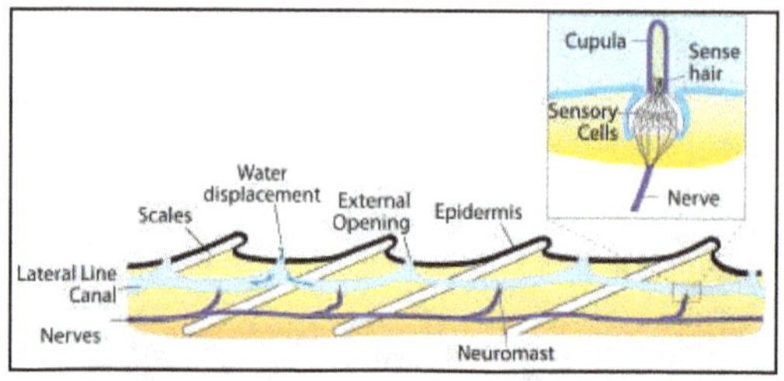

Some scales of the lateral line (center) of a *Rutilus rutilus*.

Hair cells utilize a system of transduction that uses rate coding in order to transmit the directionality of a stimulus. Hair cells of the lateral line system produce a constant, tonic rate of firing. As mechanical motion is transmitted through water to the neuromast, the cupula bends and is displaced. Varying in magnitude with the strength of the stimulus, shearing movement and deflection of the hairs is produced, either toward the longest hair or away from it. This results in a shift in the cell's ionic permeability, resulting from changes to open ion channels caused by the deflection of the hairs. Deflection towards the longest hair results in depolarization of the hair cell, increased neurotransmitter release at the excitatory afferent synapse, and a higher rate of signal transduction. Deflection towards the shorter hair has the opposite effect, hyperpolarizing the hair cell and producing a decreased rate of neurotransmitter release. These electrical impulses are then transmitted along afferent lateral neurons to the brain.

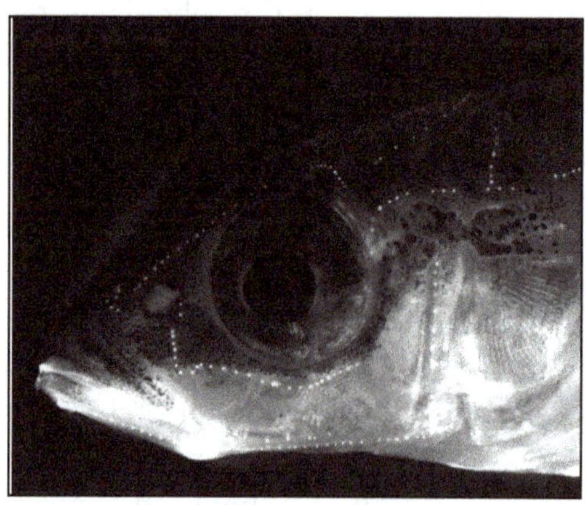

A *three-spined stickleback* with stained neuromasts.

While both varieties of neuromasts utilize this method of transduction, the specialized organization of superficial and canal neuromasts allow them different mechanoreceptive capacities. Located at the surface of an animal's skin, superficial organs are exposed more directly to the external environment. Though these organs possess the standard "staircase" shaped hair bundles, overall the organization of the bundles within the organs is seemingly haphazard, incorporating various shapes and sizes of microvilli within bundles. This suggests a wide range of detection, potentially indicating a function of broad detection to determine the presence and magnitude of deflection caused by motion in the surrounding water. In contrast, the structure of canal organs allow canal neuromasts to be organized into a network system that allows more sophisticated mechanoreception, such as the detection of pressure differentials. As current moves across the pores of a canal, a pressure differential is created over the pores. As pressure on one pore exceeds that of another pore, the differential pushes down on the canal and causes flow in the canal fluid. This moves the cupula of the hair cells in the canal, resulting in a directional deflection of the hairs corresponding to the direction of the flow. This method allows the translation of pressure information into directional deflections which can be received and transduced by hair cells.

Electrophysiology

The mechanoreceptive hair cells of the lateral line structure are integrated into more complex circuits through their afferent and efferent connections. The synapses that directly participate in the transduction of mechanical information are excitatory afferent connections that utilize glutamate. However, a variety of different neuromast and afferent connections are possible, resulting in variation in mechanoreceptive properties. For instance, a series of experiments on the superficial neuromasts of *Porichthys notatus* revealed that neuromasts can exhibit a receptive specificity for particular frequencies of stimulation. Using an immobilized fish to prevent extraneous stimulation, a metal ball was vibrated at different frequencies. Utilizing single cell measurements with a microelectrode, responses were recorded and used to construct tuning curves, which revealed frequency preferences and two main afferent nerve types. One variety is attuned to collect mechanoreceptive information about acceleration, responding to stimulation frequencies between 30–200 Hz. The other type is sensitive to velocity information and is most receptive to stimulation below <30 Hz. This suggests a more intricate model of reception than was previously considered.

The efferent synapses to hair cells are inhibitory and utilize acetylcholine as a transmitter. They are crucial participants in a corollary discharge system designed to limit self-generated interference. When a fish moves, it creates disturbances in the water that could be detected by the lateral line system, potentially interfering with the detection of other biologically relevant signals. To prevent this, an afferent signal is sent to the hair cell upon motor action, resulting in inhibition which counteracts the excitation resulting from reception of the self-generated stimulation. This allows the fish to retain perception of motion stimuli without interference created by its own movements.

After signals transduced from the hair cells are transmitted along lateral neurons, they eventually reach the brain. Visualization methods have revealed that the area where these signals most often terminate is the medial octavolateralis nucleus (MON). It is likely that the MON plays an important role in the processing and integration of mechanoreceptive information. This has been supported through other experiments, such as the use of Golgi staining and microscopy by New & Coombs to demonstrate the presence of distinct cell layers within the MON. Distinct layers of basilar and non-basilar crest cells were identified within the deep MON. Drawing a comparison to similar cells in the closely related electrosensory lateral line lobe of electric fish, it seems to suggest possible computational pathways of the MON. The MON is likely involved in the integration of sophisticated excitatory and inhibitory parallel circuits in order to interpret mechanoreceptive information.

Sensory Systems in Fish

Most fish possess highly developed sense organs. Nearly all daylight fish have color vision that is at least as good as a human's. Many fish also have chemoreceptors that are responsible for extraordinary senses of taste and smell. Although they have ears, many fish may not hear very well. Most fish have sensitive receptors that form the lateral line system, which detects gentle currents and vibrations, and senses the motion of nearby fish and prey. Sharks can sense frequencies in the range of 25 to 50 Hz through their lateral line.

Fish orient themselves using landmarks and may use mental maps based on multiple landmarks or symbols. Fish behavior in mazes reveals that they possess spatial memory and visual discrimination.

Vision

Fish have a refractive index gradient within the lens of their eyes which compensates for spherical aberration. Unlike humans, most fish adjust focus by moving the lens closer or further from the retina.

Vision is an important sensory system for most species of fish. Fish eyes are similar to those of terrestrial vertebrates like birds and mammals, but have a more spherical lens. Their retinas generally have both rod cells and cone cells (for scotopic and photopic vision), and most species have colour vision. Some fish can see ultraviolet and some can see polarized light. Amongst jawless fish, the lamprey has well-developed eyes, while the hagfish has only primitive eyespots. Fish vision shows adaptation to their visual environment, for example deep sea fishes have eyes suited to the dark environment.

Fish and other aquatic animals live in a different light environment than terrestrial species. Water absorbs light so that with increasing depth the amount of light available decreases quickly. The optic properties of water also lead to different wavelengths of light being absorbed to different degrees, for example light of long wavelengths (e.g. red, orange) is absorbed quite quickly compared to light of short wavelengths (blue, violet), though ultraviolet light (even shorter wavelength than blue) is absorbed quite quickly as well. Besides these universal qualities of water, different bodies of water may absorb light of different wavelengths because of salts and other chemicals in the water.

Hearing

Hearing is an important sensory system for most species of fish. Hearing threshold and the ability to localize sound sources are reduced underwater, in which the speed of sound is faster than in air. Underwater hearing is by bone conduction, and localization of sound appears to depend on differences in amplitude detected by bone conduction. Aquatic animals such as fish, however, have a more specialized hearing apparatus that is effective underwater.

Fish can sense sound through their lateral lines and their otoliths (ears). Some fishes, such as some species of carp and herring, hear through their swim bladders, which function rather like a hearing aid.

Hearing is well-developed in carp, which have the Weberian organ, three specialized vertebral processes that transfer vibrations in the swim bladder to the inner ear.

Although it is hard to test sharks' hearing, they may have a sharp sense of hearing and can possibly hear prey many miles away. A small opening on each side of their heads (not the spiracle) leads directly into the inner ear through a thin channel. The lateral line shows a similar arrangement, and is open to the environment via a series of openings called lateral line pores. This is a reminder of the common origin of these two vibration- and sound-detecting organs that are grouped together as the acoustico-lateralis system. In bony fish and tetrapods the external opening into the inner ear has been lost.

Current Detection

Hair cells in fish are used to detect water movements around their bodies. These hair cells are embedded in a jelly-like protrusion called cupula. The hair cells therefore can not be seen and do not appear on the surface of skin.

The lateral line in fish and aquatic forms of amphibians is a detection system of water currents, consisting mostly of vortices. The lateral line is also sensitive to low-frequency vibrations. The mechanoreceptors are hair cells, the same mechanoreceptors for vestibular sense and hearing. It is used primarily for navigation, hunting, and schooling. The receptors of the electrical sense are modified hair cells of the lateral line system.

Fish and some aquatic amphibians detect hydrodynamic stimuli via a lateral line. This system consists of an array of sensors called neuromasts along the length of the fish's body. Neuromasts can be free-standing (superficial neuromasts) or within fluid-filled canals (canal neuromasts). The sensory cells within neuromasts are polarized hair cells contained within a gelatinous cupula. The cupula, and the stereocilia within, are moved by a certain amount depending on the movement of the surrounding water. Afferent nerve fibers are excited or inhibited depending on whether the hair cells they arise from are deflected in the preferred or opposite direction. Lateral line receptors form somatotopic maps within the brain informing the fish of amplitude and direction of flow at different points along the body. These maps are located in the medial octavolateral nucleus (MON) of the medulla and in higher areas such as the torus semicircularis.

Pressure Detection

Pressure detection uses the organ of Weber, a system consisting of three appendages of vertebrae transferring changes in shape of the gas bladder to the middle ear. It can be used to regulate the buoyancy of the fish. Fish like the weather fish and other loaches are also known to respond to low pressure areas but they lack a swim bladder.

Chemoreception

The shape of the hammerhead shark's head may enhance olfaction by spacing the nostrils further apart.

Sharks have keen olfactory senses, located in the short duct (which is not fused, unlike bony fish) between the anterior and posterior nasal openings, with some species able to detect as little as one part per million of blood in seawater.

Sharks have the ability to determine the direction of a given scent based on the timing of scent detection in each nostril. This is similar to the method mammals use to determine direction of sound.

They are more attracted to the chemicals found in the intestines of many species, and as a result often linger near or in sewage outfalls. Some species, such as nurse sharks, have external barbels that greatly increase their ability to sense prey.

The MHC genes are a group of genes present in many animals and important for the immune system; in general, offspring from parents with differing MHC genes have a stronger immune system. Fish are able to smell some aspect of the MHC genes of potential sex partners and prefer partners with MHC genes different from their own.

Salmon have a strong sense of smell. Speculation about whether odours provide homing cues, go back to the 19th century. In 1951, Hasler hypothesised that, once in vicinity of the estuary or entrance to its birth river, salmon may use chemical cues which they can smell, and which are unique to their natal stream, as a mechanism to home onto the entrance of the stream. In 1978, Hasler and his students convincingly showed that the way salmon locate their home rivers with such precision was indeed because they could recognise its characteristic smell. They further demonstrated that the smell of their river becomes imprinted in salmon when they transform into smolts, just before they migrate out to sea. Homecoming salmon can also recognise characteristic smells in tributary streams as they move up the main river. They may also be sensitive to characteristic pheromones given off by juvenile conspecifics. There is evidence that they can "discriminate between two populations of their own species".

Electroreception and Magnetoreception

Electroreception, or electroception, is the ability to detect electric fields or currents. Some fish, such as catfish and sharks, have organs that detect weak electric potentials on the order of millivolts. Other fish, like the South American electric fishes Gymnotiformes, can produce weak electric currents, which they use in navigation and social communication. In sharks, the ampullae of

Lorenzini are electroreceptor organs. They number in the hundreds to thousands. Sharks use the ampullae of Lorenzini to detect the electromagnetic fields that all living things produce. This helps sharks (particularly the hammerhead shark) find prey. The shark has the greatest electrical sensitivity of any animal. Sharks find prey hidden in sand by detecting the electric fields they produce. Ocean currents moving in the magnetic field of the Earth also generate electric fields that sharks can use for orientation and possibly navigation.

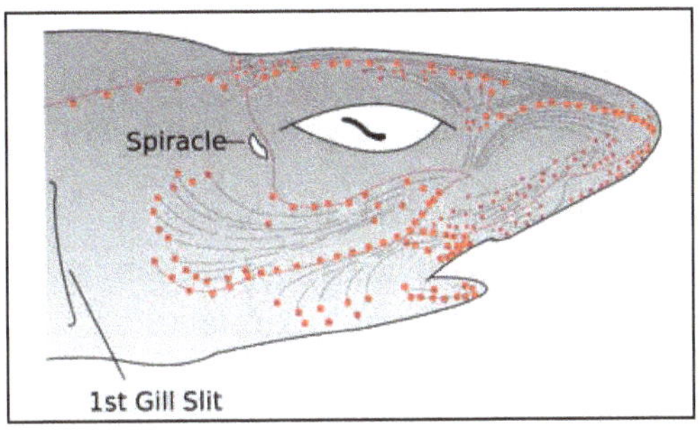

Electromagnetic field receptors (ampullae of Lorenzini) and motion detecting canals in the head of a shark.

Electric field proximity sensing is used by the electric catfish to navigate through muddy waters. These fish make use of spectral changes and amplitude modulation to determine factors such shape, size, distance, velocity, and conductivity. The abilities of the electric fish to communicate and identify sex, age, and hierarchy within the species are also made possible through electric fields. EF gradients as low as 5nV/cm can be found in some saltwater weakly electric fish.

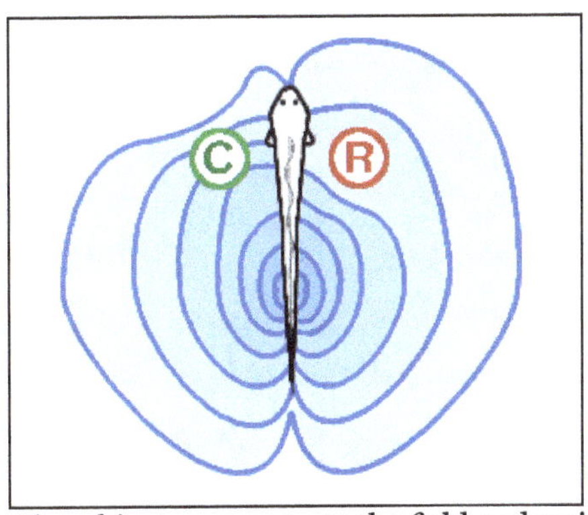

Active electrolocation. Conductive objects concentrate the field and resistive objects spread the field.

The paddlefish (*Polyodon spathula*) hunts plankton using thousands of tiny passive electroreceptors located on its extended snout, or rostrum. The paddlefish is able to detect electric fields that oscillate at 0.5–20 Hz, and large groups of plankton generate this type of signal.

Electric fishes use an active sensory system to probe the environment and create active electrodynamic imaging.

In 1973, it was shown that Atlantic salmon have conditioned cardiac responses to electric fields with strengths similar to those found in oceans. "This sensitivity might allow a migrating fish to align itself upstream or downstream in an ocean current in the absence of fixed references."

Magnetoception, or magnetoreception, is the ability to detect the direction one is facing based on the Earth's magnetic field. In 1988, researchers found iron, in the form of single domain magnetite, resides in the skulls of sockeye salmon. The quantities present are sufficient for magnetoception.

Fish Navigation

Salmon regularly migrate thousands of miles to and from their breeding grounds.

Salmon spend their early life in rivers, and then swim out to sea where they live their adult lives and gain most of their body mass. When they have matured, they return to the rivers to spawn. Usually they return with uncanny precision to the natal river where they were born, and even to the very spawning ground of their birth. It is thought that, when they are in the ocean, they use magnetoception to locate the general position of their natal river, and once close to the river, that they use their sense of smell to home in on the river entrance and even their natal spawning ground.

After several years wandering huge distances in the ocean, most surviving salmon return to the same natal rivers where they were spawned. Then most of them swim up the rivers until they reach the very spawning ground that was their original birthplace.

There are various theories about how this happens. One theory is that there are geomagnetic and chemical cues which the salmon use to guide them back to their birthplace. The fish may be sensitive to the Earth's magnetic field, which could allow the fish to orient itself in the ocean, so it can navigate back to the estuary of its natal stream.

Pain

Hooked sailfish

Experiments done by William Tavolga provide evidence that fish have pain and fear responses. For instance, in Tavolga's experiments, toadfish grunted when electrically shocked and over time they came to grunt at the mere sight of an electrode.

In 2003, Scottish scientists at the University of Edinburgh and the Roslin Institute concluded that rainbow trout exhibit behaviors often associated with pain in other animals. Bee venom and acetic acid injected into the lips resulted in fish rocking their bodies and rubbing their lips along the sides and floors of their tanks, which the researchers concluded were attempts to relieve pain, similar to what mammals would do. Neurons fired in a pattern resembling human neuronal patterns.

Professor James D. Rose of the University of Wyoming claimed the study was flawed since it did not provide proof that fish possess "conscious awareness, particularly a kind of awareness that is meaningfully like ours". Rose argues that since fish brains are so different from human brains, fish are probably not conscious in the manner humans are, so that reactions similar to human reactions to pain instead have other causes. Rose had published a study a year earlier arguing that fish cannot feel pain because their brains lack a neocortex. However, animal behaviorist Temple Grandin argues that fish could still have consciousness without a neocortex because "different species can use different brain structures and systems to handle the same functions."

Animal welfare advocates raise concerns about the possible suffering of fish caused by angling. Some countries, such as Germany have banned specific types of fishing, and the British RSPCA now prosecutes individuals who are cruel to fish.

Sensory Organs of Gastropods

The sensory organs of gastropods (snails and slugs) include olfactory organs, eyes, statocysts and mechanoreceptors. Gastropods have no sense of hearing.

Olfactory Organs

Rhinophores of *Aplysia californica*.

In terrestrial gastropods the most important sensory organs are the olfactory organs which are located on the tips of the 4 tentacles.

In opisthobranch marine gastropods, the chemosensory organs are two protruding structures on top of the head. These are known as rhinophores.

The upper pair of tentacles on the head of the edible snail *Helix pomatia* have eyes, but the main sensory organs are sensory neurons for olfaction in the epithelium of the tentacles.

An opisthobranch sea slug *Navanax inermis* has chemoreceptors on the sides of its mouth to track mucopolysaccharides in the slime trails of prey, and of potential mates.

The freshwater snail *Bithynia tentaculata* is capable of detecting the presence of molluscivorous (mollusk-eating) leeches through chemoreception, and of closing its operculum to avoid predation.

The deepwater snail *Bathynerita naticoidea* can detect mussel beds containing the mussel *Bathymodiolus childressi*, because it is attracted to water that has cues in it from this species of mussel.

Some terrestrial gastropods can track the odor of food using their tentacles (tropotaxis) and the wind (anemotaxis).

Eyes

In terrestrial pulmonate gastropods, eye spots are present at the tips of the tentacles in the Stylommatophora or at the base of the tentacles in the Basommatophora. These eye spots range from simple ocelli that cannot project an image (simply distinguishing light and dark), to more complex pit and even lens eyes. Vision is not the most important requirement in terrestrial gastropods, because they are mainly nocturnal animals.

Some gastropods, for example the freshwater Apple snails (family Ampullariidae) and marine species of genus *Strombus* can completely regenerate their eyes. The gastropods in both of these families have lens eyes.

Morphological sequence of different types of multicellular eyes exemplified by gastropod eyes:

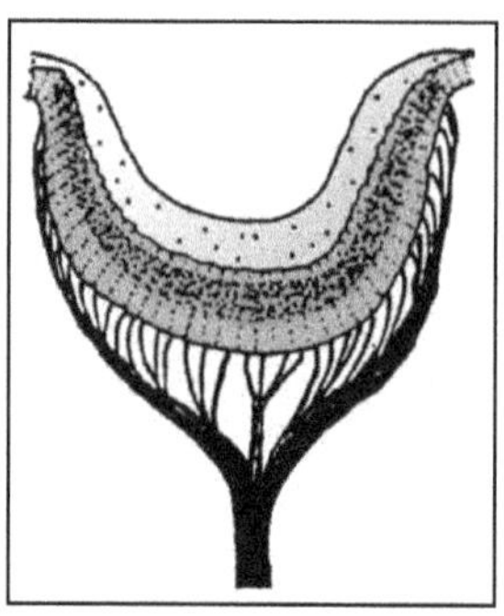

Eye pit of *patella* sp.

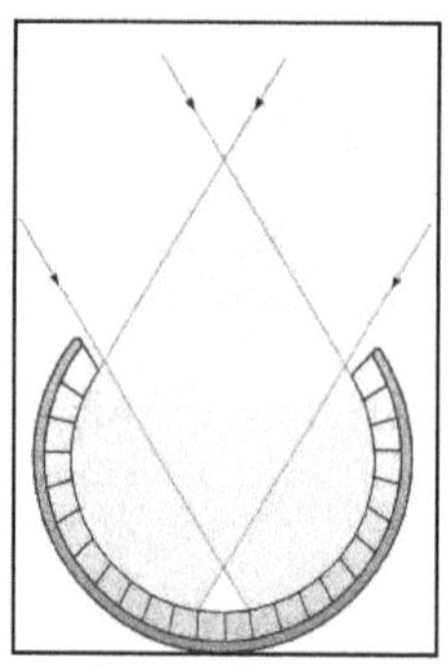

Scheme of pit eye.

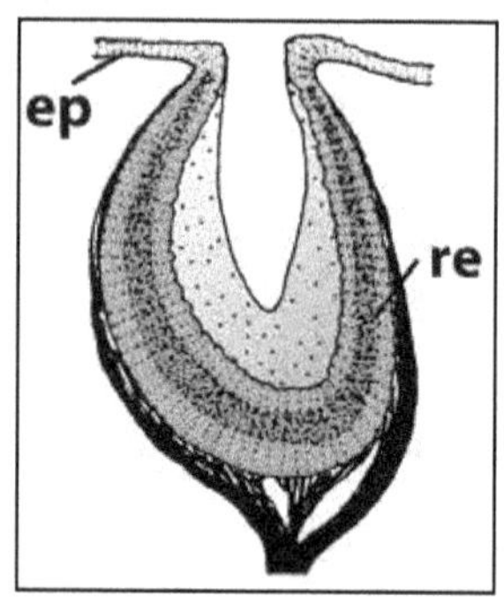

Eye cup of *Pleurotomaria* sp.

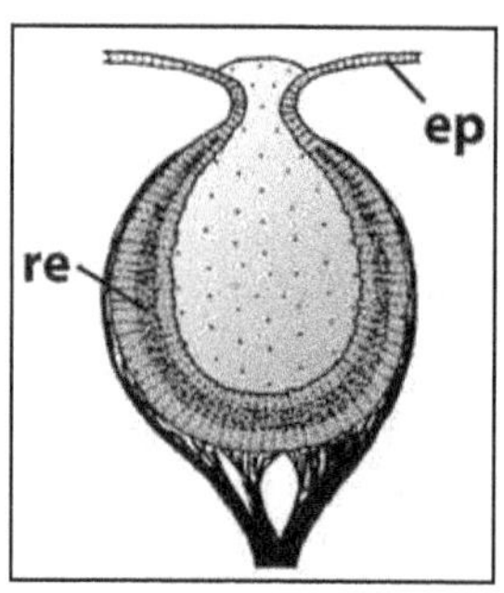

Pinhole eye of *Haliotis* sp.

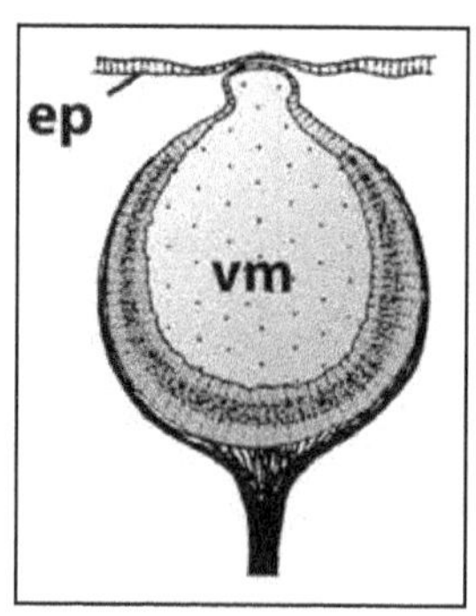

Closed eye of *Turbo coronatus*.

Lens Eyes

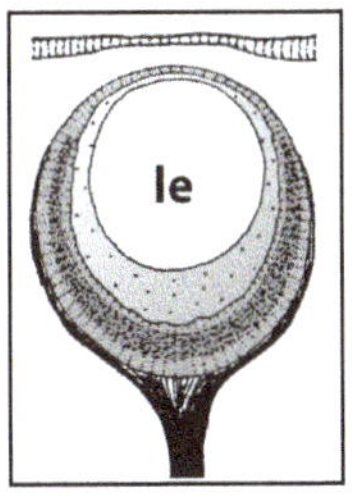

Lens eye of *Bolinus brandaris.*

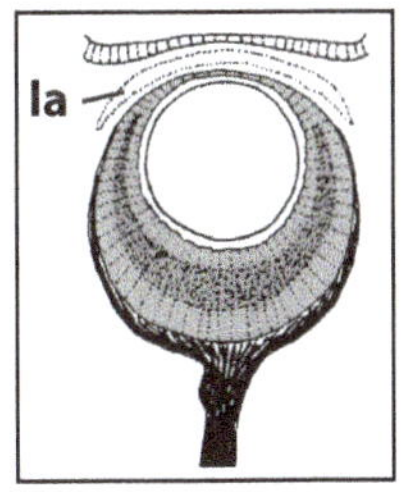

Lens eye of *Nucella lapillus.*

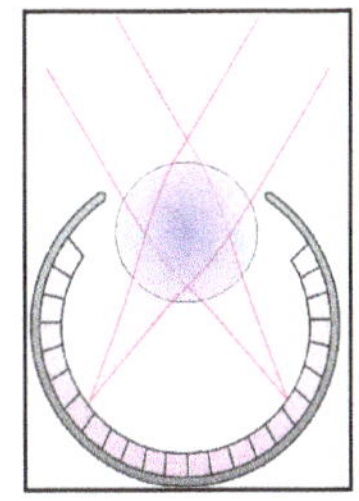

Scheme of lens eye.

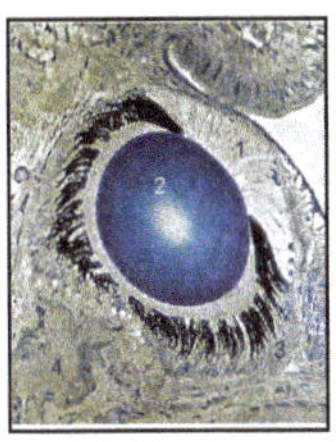

Eye of a snail.

1 - anterior chamber,

2 - lens,

3 - retina,

4 - optic nerve.

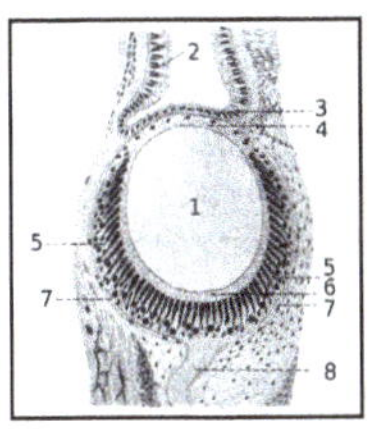

Drawing of cross section of the eye of *Helix pomatia.*

1 - lens

2 - olfactory epithelium

3 - corneal epithelium

4 - corneal endothelium

5 - retina

6 - layer with rod cells

7 - fibrous connective tissue layer

8 - nerve of the eye

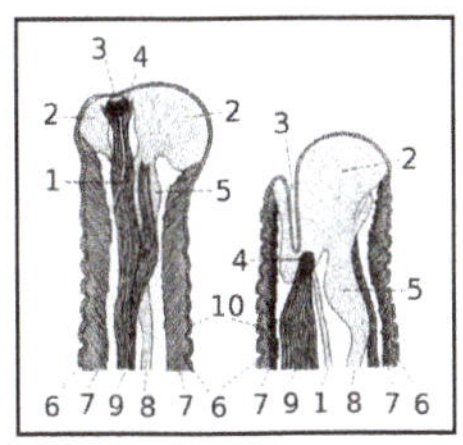

Drawing of cross sections of the extracted tentacle (left) and constricted tentacle (right) with and eye of *Helix pomatia.*

1 - nerve of an eye

4 - eye

5 - tentacle ganglion

6 - epidermis

8 - nerve of an tentacle

9 - retractor muscle

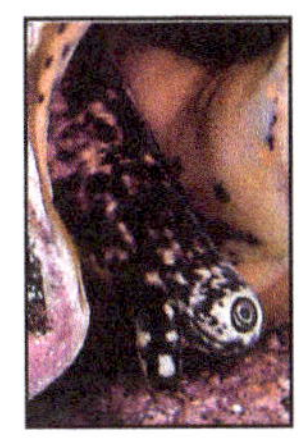

Well-developed lens eye of *Eustrombus gigas* on eyestalk has a black iris. There is a small tentacle on the eyestalk also.

Statocysts

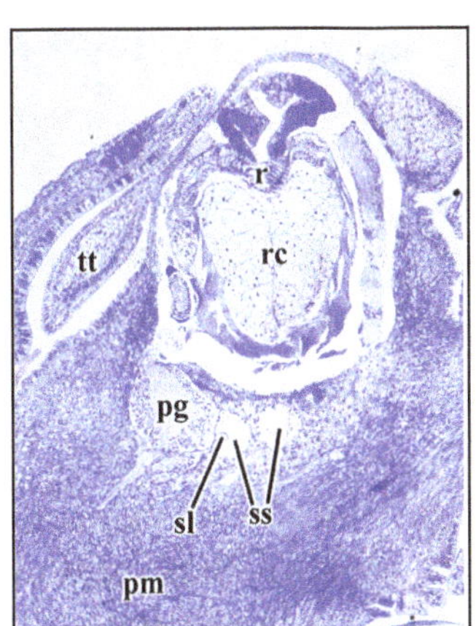

Statocysts (ss) and statolith (sl) inside the head of *Gigantopelta chessoia.*

In the statocysts of *Haliotis asinina* was found the expression of a conserved gene (Pax-258 gene), which is also important for forming structures for balance in eumetazoans.

Ampullae of Lorenzini

The ampullae of Lorenzini are special sensing organs called electroreceptors, forming a network of jelly-filled pores. They are mostly discussed as being found in cartilaginous fish (sharks, rays, and chimaeras); however, they are also reported to be found in Chondrostei such as reedfish and sturgeon. Lungfish have also been reported to have them. Teleosts have re-evolved a different type of electroreceptors. They were first described by Stefano Lorenzini in 1678.

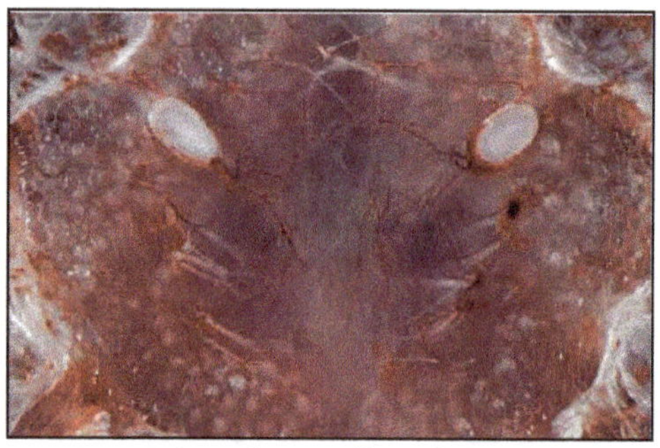

Inner view of Ampullae of Lorenzini.

These sensory organs help fish to sense electric fields in the water. Each ampulla consists of a jelly-filled canal opening to the surface by a pore in the skin and ending blindly in a cluster of small pockets full of special jelly. The ampullae are mostly clustered into groups inside the body, each cluster having ampullae connecting with different parts of the skin, but preserving a left-right symmetry. The canal lengths vary from animal to animal, but the distribution of the pores is generally specific to each species. The ampullae pores are plainly visible as dark spots in the skin. They provide fish with an additional sense capable of detecting electric and magnetic fields as well as temperature gradients.

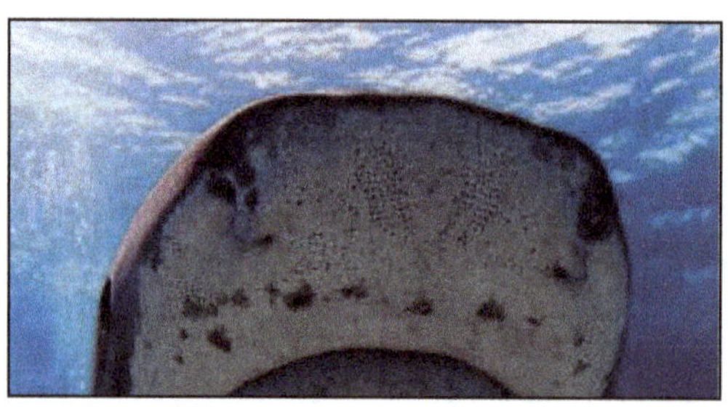

Pores with ampullae of Lorenzini in snout of Tiger shark.

Electric Field Sensing Ability

The ampullae detect electric fields in the water, or more precisely the potential difference between the voltage at the skin pore and the voltage at the base of the electroreceptor cells. A positive pore stimulus would decrease the rate of nerve activity coming from the electroreceptor cells, and a negative pore stimulus would increase the rate of nerve activity coming from the electroreceptor cells. Each ampulla contains a single layer of cells that contains electrically excitable receptor cells separated by supporting cells. The cells are connected by apical tight junctions so that no current leaks between the cells. The apical faces of the receptor cells have a small surface area with a high concentration of voltage dependent calcium channels and calcium activated potassium channels. Because the canal wall has a very high resistance, all of the voltage difference between the pore of the canal and the ampulla is dropped across the receptor epithelium which is about 50 microns thick. Because the basal membranes of the receptor cells have a lower resistance, most of the voltage is dropped across the apical faces which are excitable and are poised at threshold. Inward calcium current across the receptor cells depolarizes the basal faces causing presynaptic calcium release and release of excitatory transmitter onto the afferent nerve fibers. One of the first descriptions of calcium activated potassium channels was based on studies of the ampulla of Lorenzini in the skate. Large conductance calcium activated potassium channels (BK channels) have recently been demonstrated in the ampulla by cloning.

Sharks may be more sensitive to electric fields than any other animal, with a threshold of sensitivity as low as 5 nV/cm. That is 5/1,000,000,000 of a volt measured in a centimeter-long ampulla. All living creatures produce an electrical field by muscle contractions, and a shark may pick up weak electrical stimuli from the muscle contractions of animals, particularly prey. On the other hand, the electrochemical fields generated by paralyzed prey were sufficient to elicit a feeding attack from sharks and rays in experimental tanks; therefore muscle contractions are not necessary to attract the animals. Sharks and rays can locate prey buried in the sand, or DC electric dipoles that simulate the main feature of the electric field of a prey buried in the sand.

Any moving conductor, such as sea water, induces an electric field when a magnetic field, such as the earth's is present. The electric fields induced in oceanic currents by the Earth's magnetic field are of the same order of magnitude as the electric fields that sharks and rays are capable of sensing. This could mean that sharks and rays can orient to the electric fields of oceanic currents, and use other sources of electric fields in the ocean for local orientation. Additionally, the electric field they induce in their bodies when swimming in the magnetic field of the Earth may enable them to sense their magnetic heading.

Behavioral studies have also provided evidence that sharks can detect changes in the geomagnetic field. In one experiment, sandbar sharks and scalloped hammerhead sharks were conditioned to associate a food reward with an artificial magnetic field. When the food reward was removed, the sharks continued to show a marked difference in behavior when the magnetic field was turned on as compared to when it was off.

Temperature Sensing Ability

Early in the 20th century, the purpose of the ampullae was not clearly understood, and electrophysiological experiments suggested a sensibility to temperature, mechanical pressure and possibly salinity. It was not until 1960 that the ampullae were clearly identified as specialized receptor organs for sensing electric fields. The ampullae may also allow the shark to detect changes in water temperature. Each ampulla is a bundle of sensory cells containing multiple nerve fibres. These fibres are enclosed in a gel-filled tubule which has a direct opening to the surface through a pore. The gel is a glycoprotein based substance with the same resistivity as seawater, and it has electrical properties similar to a semiconductor. This has been suggested as a mechanism by which temperature changes are transduced into an electrical signal that the shark may use to detect temperature gradients, although it is a subject of debate in scientific literature.

Material Properties

The hydrogel, which contains keratan sulfate in 97% water, has a conductivity of about 1.8 mS/cm, the highest known amongst biological materials.

Cercus

Earwig with large cerci (top).

Cerci (singular cercus) are paired appendages on the rear-most segments of many arthropods, including insects and symphylans. Many forms of cerci serve as sensory organs, but some serve as pinching weapons or as organs of copulation. In many insects, they simply may be functionless vestigial structures.

In basal arthropods, such as silverfish, the cerci originate from the eleventh abdominal segment. As segment eleven is reduced or absent in the majority of arthropods, in such cases, the cerci emerge from the tenth abdominal segment. It is not clear that other structures so named are homologous. In the Symphyla they are associated with spinnerets.

Morphology and Functions

Most cerci are segmented and jointed, or filiform (threadlike), but some take very different forms. Some Diplura, in particular *Japyx* species, have large, stout forcipate (pincer-like) cerci that they use in capturing their prey.

The Dermaptera, or earwigs, are well known for the forcipate cerci that most of them bear, though species in the suborders Arixeniina and Hemimerina do not. It is not clear how many of the Dermaptera use their cerci for anything but defense, but some definitely feed on prey caught with the cerci, much as the Japygidae do.

Crickets have particularly long cerci while other insects have cerci that are too small to be noticeable. However, it is not always obvious that small cerci are without function; they are rich in sensory cells and may be of importance in guiding copulation and oviposition.

Some insects such as mayflies, silverfish, and bristletails have an accompanying third central tail filament which extends from the tip of the abdomen. This is referred to as the *terminal filament* and is not regarded as a cercus.

Aphids have tube-like cornicles or siphunculi that are sometimes mistaken for cerci but are not morphologically related to cerci.

Evolutionary Origin

Like many insect body parts, including mandibles, antennae and styli, cerci are thought to have evolved from what were legs on the primal insect form; a creature that may have resembled a velvet worm, Symphylan or a centipede, worm-like with one pair of limbs for each segment behind the head or anterior tagma.

Crista Acustica

Crista acustica (also Siebold's organ, or crista acoustica) is a part of the hearing organ (tibial organ) in some insects (e.g. grasshoppers, crickets, katydids). It is a collection of sensory cells that form a crest (hence the name) on top of the hollow tube (the foreleg trachea) behind the hearing membrane (tympanum) on the legs of the insect.

The crista acustica is a transition from the intermedia organ (from the midline to the periphery), together which compose the tibial hearing organ (as opposed to the tympanal hearing organ).

The crista acustica is one of three fiddle-string-like organs or chordotonal organ in insects: the others are the intermediate organ and the supratympanal organ/subgenual organ. These chordotonal organs are actually collections of sensory cells sensitive to vibration (these cells are called scolopidia cells). Their cells are attached to the tube in the legs of the insects (the trachae, *"trah-key-ah"*). So when the tube vibrates, the cells vibrate. In the crista acustica, it turns out that different scolopidia cells are sensitive to different vibrations depending on the frequency of the vibration. Since these organs are sensitive to vibrations (due to changes in pressure. It turns out the cells closest to midline are the largest and sensitive to the lowest frequency (low frequencies having the largest wavelength), and the cells further out (distal) are smaller and sensitive to higher frequencies (since high frequencies have shorter wavelengths). This orderly arrangement of sensory cells gives the insect the ability to discriminate frequencies, much like the inner ear of mammals.

Forked Tongue

A forked tongue is a tongue split into two distinct tines at the tip; this is a feature common to many species of reptiles. Reptiles smell using the tip of their tongue, and a forked tongue allows them to sense from which direction a smell is coming. Sensing from both sides of the head and following trails based on chemical cues is called tropotaxis. It is unclear whether forked-tongued reptiles can actually follow trails or if this is just a hypothesis.

Forked tongue of a Carpet Python (*Morelia spilota mcdowelli*).

Forked tongues have evolved in these Squamate reptiles (lizards and snakes) for various purposes. The advantage to having a forked tongue is that more surface area is available for the chemicals to contact and the potential for tropotaxis. The tongue is flicked out of the mouth regularly to sample the chemical environment. This form of chemical sampling allows these animals to sense non-volatile chemicals, which cannot be detected by simply using the olfactory system. This increased ability to sense chemicals has allowed for heightened abilities to identify prey, recognize kin, choose mates, locate shelters, follow trails, and more.

Forked tongues have evolved multiple times in Squamates. It is unclear, based on the morphological and genetic evidence, where the exact points of change are from a notched tongue to a forked tongue, but it is believed that the change has happened two to four times. A common behavioral characteristic that has evolved in those with forked tongues is that they tend to be wide foragers.

Hummingbirds also have tongues that split at the tip. Galagos (bushbabies) have a secondary tongue, or sublingua, used for grooming, hidden under their first.

Usage as First Nations Cultural Term

The phrase "speaks with a forked tongue" means to deliberately say one thing and mean another or, to be hypocritical, or act in a duplicitous manner. In the longstanding tradition of many Native American tribes, "speaking with a forked tongue" has meant lying, and a person was no longer considered worthy of trust, once he had been shown to "speak with a forked tongue". This phrase was also adopted by Americans around the time of the Revolution, and may be found in abundant references from the early 19th century — often reporting on American officers who sought to convince the tribal leaders with whom they negotiated that they "spoke with a straight and not with a forked tongue" (as for example, President Andrew Jackson told the Creek Nation in 1829). According to one 1859 account, the native proverb that the "white man spoke with a forked tongue" originated as a result of the French tactic of the 1690s, in their war with the Iroquois, of inviting their enemies to attend a Peace Conference, only to be slaughtered or captured.

Literary Usage

There are appearances of the phrase "forked tongue" in English literature, either in reference to actual snakes' tongues, or as a metaphor for untruthfulness, such as a sermon by Lancelot Andrewes, who died in 1626:

"And he hath the art of cleaving. He shewed it in the beginning, when he made the Serpent, lingnam bisulcam, a forked tongue, to speake that, which was contrary to his knowledge and meaning, They should not die; and as hee did the Serpents, so hee can doe others."

The phrase also appears in Milton's Paradise Lost:

"According to his Doom: He would have spoke, But Hiss for Hiss return'd with forked Tongue To forked Tongue, for now were all transform'd".

Johnston's Organ

Johnston's organ is a collection of sensory cells found in the pedicel (the second segment) of the antennae in the Class Insecta. Johnston's organ detects motion in the flagellum (third and typically final antennal segment). It consists of scolopidia arrayed in a bowl shape, each of which contains a mechanosensory chordotonal neuron. The number of scolopidia varies between species. In homopterans, the Johnston's organs contain 25 - 79 scolopidia. The presence of Johnston's organ is a defining characteristic which separates the Class Insecta from the other hexapods belonging to the group Entognatha. Johnston's organ was named after the physician Christopher Johnston, father of the physician and Assyriologist Christopher Johnston.

Uses of the Johnston's Organ

In Fruit Flies

In the fruit fly *Drosophila melanogaster*, the Johnston's organ contains almost 480 sensory neurons. Distinct populations of neurons are activated differently by deflections of antennae caused by gravity or by vibrations caused by sound or air movement. This differential response allows the fly to distinguish between gravitational, mechanical, and acoustic stimuli.

The Johnston's organ of fruit flies can be used to detect air vibrations caused by the wingbeat frequency or courtship song of a mate. One function of the Johnston's organ is for detecting the wing beat frequency of a mate. Production of sound in air results in two energy components: the pressure component, which is changes in air pressure emanating away from the source of the sound; and the particle displacement component, which is the back and forth vibration of air particles oscillating in the direction of sound propagation. Particle displacement has greater energy loss than the pressure component, so the displacement component, called "near-field sound," is detectable only within one wavelength of the source.

Insects, such as fruit flies and bees, detect near field sounds using loosely attached hairs or antennae which vibrate with air particle movement. (Tympanal organs detect the pressure component of sound.) Near-field sound, because of the rapid dissipation of energy, is suitable only for very close communication. Two examples of near-field sound communication are bee's waggle dance and Drosophila courtship songs. In fruit flies, the arista of the antennae and the third segment act as the sound receiver. Vibrations of the receiver cause rotation of the third segment, which channels sound input to the mechanoreceptors of the Johnston's Organ.

In Hawk Moths

The Johnston's organ plays a role in the control of flight stability in hawk moths. Kinematic data measured from hovering moths during steady flight indicate that the antennae vibrate with a frequency matching wingbeat (27 Hz). During complex flight, however, angular changes of the flying moth cause coriolis forces, which are predicted to manifest as a vibration of the antenna of at about twice wingbeat frequency (~60 Hz). When antennae were manipulated to vibrate at a range of frequencies and the resulting signals from the neurons associated with the Johnston's organs were measured, the response of the scolopidia neurons to the frequency was tightly coupled in the range of 50–70 Hz, which is the predicted range of vibrations caused by Coriolis effects. Thus, the Johnston's organ is tuned to detect angular changes during maneuvering in complex flight.

In Honeybees

Dancing honeybees (*Apis mellifera*) describe the location of nearby food sources by emitted airborne sound signals. These signals consist of rhythmic high-velocity movement of air particles. These near field sounds are received and interpreted using the Johnston's organ in the pedicel of the antennae. Honeybees also perceive electric field changes via the Johnston's organs in their antennae and possibly other mechano-receptors. Electric fields generated by movements of the wings cause displacements of the antennae based on Coulomb's law. Neurons of the Johnston's organ respond to movements within the range of displacements caused by electric fields. When the antennae were prevented from moving at the joints containing the Johnston's organ, bees no longer responded to biologically relevant electric fields. Honeybees respond differently to different temporal patterns. Honeybees appear to use the electric field emanating from the dancing bee for distance communication.

Mauthner Cell

The Mauthner cells are a pair of big and easily identifiable neurons (one for each half of the body) located in the rhombomere 4 of the hindbrain in fish and amphibians that are responsible for a very fast escape reflex (in the majority of animals – a so-called C-start response). The cells are also notable for their unusual use of both chemical and electrical synapses.

Evolutionary History

Mauthner cells first appear in lampreys (being absent in hagfish and lancelets), and are present in virtually all teleost fish, as well as in amphibians (including postmetamorphic frogs and toads). Some fish, such as lumpsuckers, seem to have lost the Mauthner cells however.

Role in Behavior

The C-start

A C-start is a type of a very quick startle or escape reflex that is employed by fish and amphibians (including larval frogs and toads). There are two sequential stages in the C-start: first, the head rotates about the center of mass towards the direction of future escape, and the body of the animal exhibits a curvature that resembles a letter C; then, at the second stage, the animal is propelled forward. The duration of these stages varies from species to species from about 10 to 20 ms for the first stage, and from 20 to 30 ms for the second. In fish this forward propulsion does not require contraction of the antagonistic muscle, but results from the body stiffness and the hydrodynamic resistance of the tail. When an antagonistic muscular contraction does occur during stage 2, the fish rotates in the opposite direction, producing a counter-turn, and a directional change.

The Role of the Mauthner Cell In The C-start Behavior

In cases when an abrupt acoustic, tactile or visual stimulus elicits a single action potential in one M-cell, it always correlates with a contralateral C-start escape. An extremely quick mutual feedback inhibitory circuit then assures that only one M-cell reaches spiking threshold—as the C-start has to be unilateral by definition—and that only one action potential is fired.

The Mauthner cell-mediated C-start reflex is very quick, with about 5-10 ms latency between the acoustic/tactile stimulus and the Mauthner cell discharge, and only about 2 ms between the discharge and the unilateral muscle contraction. Mauthner cells are thus the quickest motor neuron to respond to the stimulus. It makes the C-start response behaviorally important as a way to initiate the escape reflex in an *all or nothing* fashion, while the direction and speed of the escape can be corrected later through the activity of smaller motor neurons.

In larval zebrafish about ~60% of the total population of reticulospinal neurons are also activated by a stimulus that elicits the M-spike and C-start escape. A well-studied group of these reticulospinal neurons are the bilaterally paired M-cell homologues denoted MiD2cm and MiD3cm. These neurons exhibit morphological similarities to the M-cell including a lateral and ventral dendrite. They are located in rhombomeres 5 and 6 of hindbrain respectively, and also receive auditory input in parallel with the M-cell from the pVIIIth nerve. In fish, water jet stimuli that activate these neurons elicit non-mauthner initiated C-starts of a longer latency, compared with M-cell associated ones.

Although the M-cell is often considered the prototype of a command neuron in vertebrates, this designation may not be fully warranted. Although electrical stimulation of the M-cell is sufficient for eliciting a C-start, this C-start is normally weaker than the one evoked by a sensory stimulus. Moreover, the C-start can be evoked even with the M-cell ablated, although in this case the latency of the response increases. The most widely accepted model of the M-cell system, or brainstem escape network, is that the M-cell initiates a fixed action pattern to the left or right by activating a spinal motor circuit initially described by J. Diamond and colleagues, but the precise trajectory of the escape is encoded by population activity in the other classes of reticulospinal neurons functioning in parallel to the M-cell. This notion is supported by studies using *in vivo* calcium imaging in larval zebrafish which show that MiD2cm and MiD3cm are activated along with the M-cell when an offending stimulus is directed towards the head but not the tail, and are correlated with C-starts of a larger initial turn angle.

Another component of the escape response is mediated by cranial relay neurons that are activated by the Mauthner cell spike. These neurons are electrically coupled with motoneurons which innervate extraocular, jaw and opercular muscles and mediate pectoral fin adduction in hatchetfish. This component of the neural circuit was first described by Michael V.L. Bennett and colleagues.

Mauthner Cells in other Types of Behavior

Mauthner cells may be involved into behavioral patterns other than the C-start, if these types of behavior also require extremely quick bending movement of the body. Thus in goldfish Mauthner cells are activated during prey capture near the surface of the water, as this type of hunting is dangerous for the fish, and it would benefit from leaving the surface as soon as possible after the prey is captured.

In adult postmetamorphic anurans (frogs and toads) that do not have a tail, M-cells are nevertheless preserved and their discharges are associated with rapid movement of legs during an escape.

Morphology and Connections

Inputs to the M-cell: Excitation and Feed Forward Inhibition

The M-cell has two primary aspiny (lacking dendritic spines) dendrites which receive segregated inputs from various parts of the neural system. One dendrite projects laterally and the other projects either in the ventral or medial direction, depending on the species.

The ventral dendrite receives information from the optic tectum and spinal cord while the lateral dendrite receives inputs from the octovolateralis systems (the lateral line, acoustic inputs from the inner ear, and inertial information from the statoliths brought by the cranial nerve VIII).

The fibers from the ipsilateral cranial nerve VIII terminate in excitatory mixed electrical and glutamatergic synapses on the M-cell. They also electrically activate glycinergic inhibitory interneurons that terminate on the M-cells. Despite the inhibitory input having one more synapse in its pathway, there is no delay between the excitation and inhibition because the intervening synapse is electrical. It was shown that for weak stimuli the inhibition wins over the excitation, preventing the M-cell from a discharge, while for stronger stimuli excitation becomes dominant. The Inner ear afferents also terminate with electrical synapses on a population PHP inhibitory interneurons (see below) to provide an additional level of feed forward inhibition. The Mauthner cell also has GABA-, dopamine-, serotonin- and somatostatinergic inputs, each restricted to certain dendritic region.

Inputs from the optic tectum and the lateral line help control which way the C-startle bends by biasing the mauthner cells when there are obstacles in the vicinity. In cases where movement away from the stimulus is blocked, the fish may bend towards the disturbance.

Axon Cap

The Mauthner cell axon hillock is surrounded by a dense formation of neuropil, called the axon cap. The high resistance of this axon cap contributes to the typical shape of the Mauthner cell field potential. In its most advanced form the axon cap consists of a core, immediately adjacent to the Mauthner cell axon, and containing a network of very thin unmyelinated fibers, and a peripheral part. This peripheral part contains the large unmyelinated fibers of the PHP neurons that mediate the inhibitory feedback to the Mauthner cell; the Mauthner cell itself also sends small dendrites from its axon hill to the peripheral part of the axon cap. Finally, the surface of the axon cap is covered with a *cap wall* composed of several layers of astrocyte-like glial cells. Both glial cells and the unmyelinated fibers are coupled with each other by means of gap junctions.

Evolutionarily, the axon cap is a more recent development than the Mauthner cell itself, so some animals, such as lampreys and eels, while having functional Mauthner cells, don't have axon cap at all, while some other animals, such as amphibia and lungfish, do have a very simplified version of it.

Feedback Network

The main part of the Mauthner cell-associated network is the negative feedback network, which assures that only one of the two Mauthner cells fires in response to the stimulus and that, whichever Mauthner cell fires, it does so only once. Both these requirements are quite natural considering that the consequences of a single Mauthner cell discharge are so strong; a failure to comply with these two rules would not only prevent the animal from escaping, but could even physically damage it. The fastest part of this negative feedback network, which is also the one closest to the Mauthner cell, is that of the so-called *passive hyperpolarizing field potential* or PHP neurons. The fibers of these neurons are located in the axon cap, and they receive inputs from both ipsilateral and contralateral Mauthner cells. The field potentials of PHP neurons are strongly positive, and form a part of the 'Signature field potential' of the Mauthner cell, with the early (ipsilaterally initiated) component being called the Extracellular Hyperpolarizing Potential (EHP), and the later (contralaterally initiated) component being sometimes addressed in the literature as the Late Collateral Inhibition (LCI). The action of PHP neurons onto the Mauthner cells is mediated by electrical, and not chemical effects: the outward currents generated by the action potentials in axon cap fibers flow inward across the Mauthner cell axon hillock and hyperpolarize it.

Outputs

The only axon of the Mauthner cell reaches from the cell to the midline of the hindbrain, promptly crosses it to the contralateral side, and then descends caudally along the spinal cord. A single discharge of the M-cell achieves a whole set of parallel effects onto the spinal motor networks: 1) it monosynaptically excites large primary motoneurons at one side of the body; 2) disinaptically excites smaller motoneurons at the same side of the body; 3) initiates action potentials in inhibitory interneurons electrically coupled to the M-cell axon, and by their means inhibits a) inhibitory interneurons still at the same side of the body (to prevent them from interfering with the C-start), as well as b) motoneurons at the other side of the body. As a result of this pattern of activation the quick muscles at one side of the body contract simultaneously, while the muscles at the other side of the body relax.

Electrophysiology

Ephaptic Properties

The inhibition of the M-cell by the PHP cells occurs by ephaptic interactions. The inhibition is brought about without a chemical synapses or electrical synaptic coupling having low resistance gap junctions joining the cells. When the region of the PHP cell axon outside the axon cap depolarizes, the influx of positive charge into the cell through voltage gated sodium channels is accompanied by a passive outflow of current from the PHP cell axon into the region bound by the axon cap. Due to the high resistance of the surrounding glial cells, the charge does not dissipate and the potential across the M-cell membrane is increased, hyperpolarizing it.

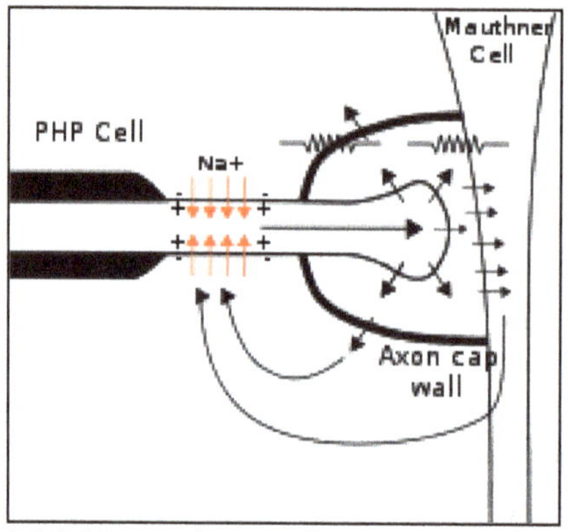

Ephaptic inhibition at the mauthner axon cap by PHP cells.

Signature Field Potential

Because of its size, presence of a quick feedback network, and abundance of electrical and quasi-electrical (ephaptic) synapses, the Mauthner cell has a strong field potential of a very characteristic shape. This field potential starts with a high-amplitude potential sink up to tens of millivolts in amplitude that originates from the Mauthner cell discharge, and which is closely followed by a positive potential, called Extrinsic Hyperpolarizing Potential or EHP, which is associated with the activity of the recurrent feedback network.

Due to its high amplitude, in some animals the negative part of Mauthner cell field potential can be detected up to several hundred micrometres away from the cell itself. The positive components of the field potential are strongest in the axon cap, reaching amplitudes of 45 mV in adult goldfish. With a knowledge of these properties of the field potential, it is possible to use field potential monitoring as a way to find the Mauthner cell body in vivo, or in vitro in a whole brain preparation, moving the recording electrode in the hindbrain, while at the same time stimulating the spinal cord, thus evoking antidromic action potentials in the Mauthner cell axon.

Plasticity

Application of serotonin was shown to increase inhibitory inputs to the M-cell, while application of dopamine – to increase the amplitude of both chemical and electrical components of the VIIIth nerve responses via a G protein-mediated activation of postsynaptic D2 receptor. An activity-dependent LTP can be evoked in M-cells by a high-frequency stimulation of the VIIIth nerve. Surprisingly, this LTP is electrical synapse-mediated, and is presumed to involve modification of the gap junction channels. A possibility of LTP induction by sensory stimuli in vivo, and the evidence for the LTP of inhibitory inputs to M-cells were also demonstrated.

Spontaneous preference in turn direction in young goldfish is correlated with one of the Mauthner cells being bigger than the other one. It is possible to change the preference of fish by raising them in conditions facilitating turns in a specific direction; this shift is accompanied by a correspondent change in M-cell sizes.

History of Research

The Mauthner cell was first identified by the Viennese ophthalmologist Ludwig Mauthner in the teleost fish for its associated neural circuit which mediates an escape response called the C-start or C-startle to direct the fish away from a predator.

The M-cell is a model system in the field of Neuroethology. The M-cell system has served for detailed neurophysiological and histological investigations of synaptic transmission and synaptic plasticity. Studies by Donald Faber and Henri Korn helped to establish the one vesicle hypothesis of synaptic transmission in the CNS. Other important research topics that have been investigated in the M-cell system include studies by Yoichi Oda and colleagues on inhibitory long-term potentiation and auditory conditioning of the startle response, and studies by Alberto Pereda and colleagues on plasticity of electrical synapses. Other research topics investigated in the M-cell system include studies of spinal neural networks and neural regeneration by Joe Fetcho and colleagues, as well as underwater sound localization, and the biophysics of computation in single neurons.

Medullary Command Nucleus

The medullary command nucleus (MCN), also called the *pacemaker nucleus,* is a group of nerve cells found in the bodies of weakly electric fish. It controls the function of electrocytes by regulating the frequency of electrical impulses. Signals originating in the MCN are transmitted to electrocytes, where changes in ion concentration cause electrical charges to be generated. The nucleus both sends and receives signals, thereby acting as a regulator and central processor for the electro sensors in the fish's body. Inputs into the MCN originate in the mesencephalic precommand nucleus, thalamic dorsal posterior nucleus, and toral ventroposterior nucleus. All of these nuclei have dense projections into the MCN, with the exception of theToral Ventroposterior nucleus, which contain only a ventral edge projection.

References

- Plachta D T T; Hanke W; Bleckmann H (2003). "A hydrodynamic topographic map in the midbrain of goldfish Carassius auratus". Journal of Experimental Biology. 206 (19): 3479–86. doi:10.1242/jeb.00582
- Shupak A. Sharoni Z. Yanir Y. Keynan Y. Alfie Y. Halpern P. (January 2005). "Underwater Hearing and Sound Localization with and without an Air Interface". Otology & Neurotology. 26 (1): 127–130. doi:10.1097/00129492-200501000-00023
- Meyer CG; Holland KN; Papastamatiou YP (2005). "Sharks can detect changes in the geomagnetic field". Journal of the Royal Society, Interface. 2 (2): 129–30. PMC 1578252. PMID 16849172. doi:10.1098/rsif.2004.0021
- Journal of Undergraduate Life Sciences. "Appropriate maze methodology to study learning in fish" (PDF). Retrieved 28 May 2009
- Boehm T; Zufall F (February 2006). "MHC peptides and the sensory evaluation of genotype". Trends in Neurosciences. 29 (2): 100–7. PMID 16337283. doi:10.1016/j.tins.2005.11.006
- Chase R.: Sensory Organs and the Nervous System. in Barker G. M. (ed.): The biology of terrestrial molluscs. CABI Publishing, Oxon, UK, 2001, ISBN 0-85199-318-4. 1-146, cited pages: 179-211
- "Rose, J.D. 2003. A Critique of the paper: "Do fish have nociceptors: Evidence for the evolution of a vertebrate sensory system"" (PDF). Retrieved 21 May 2011

- Dreller, C.; Kirchner, W. H. (1993). "Hearing in honeybees: localization of the auditory sense organ". Journal of Comparative Physiology A. 173 (3): 275–279. doi:10.1007/bf00212691
- Götting, Klaus-Jürgen (1994). "Schnecken". In Becker, U.; Ganter, S.; Just, C.; Sauermost, R. Lexikon der Biologie. Heidelberg: Spektrum Akademischer Verlag. ISBN 3-86025-156-2
- Canfield JG, Rose GJ (1993). "Activation of Mauthner neurons during prey capture". Journal of Comparative Physiology A. 172 (5): 611–618. doi:10.1007/BF00213683

Chapter 5

Locomotion and Movement of Animals

The displacement of an animal from one place to another is called locomotion. Movement is the motion of body parts of the organism without changing the location of the organism. Locomotion helps the animal find food and shelter, escape from predators or even facilitate mating. This chapter covers the different types of locomotion in animals such as arboreal locomotion, terrestrial locomotion, aquatic locomotion, passive locomotion, etc. in a comprehensive manner.

Animal Locomotion

Animal locomotion, in ethology, is any of a variety of movements or methods that animals use to move from one place to another. Some modes of locomotion are (initially) self-propelled, e.g., running, swimming, jumping, flying, hopping, soaring and gliding. There are also many animal species that depend on their environment for transportation, a type of mobility called passive locomotion, e.g., sailing (some jellyfish), kiting (spiders) and rolling (some beetles and spiders).

A cheetah chasing prey. In its brief hunting sprints, it is the fastest of all land animals.

Animals move for a variety of reasons, such as to find food, a mate, a suitable microhabitat, or to escape predators. For many animals, the ability to move is essential for survival and, as a result, natural selection has shaped the locomotion methods and mechanisms used by moving organisms. For example, migratory animals that travel vast distances (such as the Arctic tern) typically have a locomotion mechanism that costs very little energy per unit distance, whereas non-migratory animals that must frequently move quickly to escape predators are likely to have energetically costly, but very fast, locomotion.

A bee in flight.

Etymology

The term "locomotion" is formed in English from Latin *loco* "from a place" (ablative of *locus* "place") + *motio* "motion, a moving".

Locomotion in Different Media

Animals move through, or on, four types of environment: aquatic (in or on water), terrestrial (on ground or other surface, including arboreal, or tree-dwelling), fossorial (underground), and aerial (in the air). Many animals—for example semi-aquatic animals, and diving birds—regularly move through more than one type of medium. In some cases, the surface they move on facilitates their method of locomotion.

Aquatic

Swimming

Dolphins surfing at Snapper Rocks, Queensland, Australia.

In water, staying afloat is possible using buoyancy. If an animal's body is less dense than water, it can stay afloat. This requires little energy to maintain a vertical position, but requires more energy for locomotion in the horizontal plane compared to less buoyant animals. The drag encountered in water is much greater than in air. Morphology is therefore important for efficient locomotion, which is in most cases essential for basic functions such as catching prey. A fusiform, torpedo-like body form is seen in many aquatic animals, though the mechanisms they use for locomotion are diverse.

The primary means by which fish and marine mammals generate thrust is by oscillating the body, from side to side in fish, up and down in marine mammals, the resulting wave motion ending at a large tail fin or flippers. Finer control, such as for slow movements, is often achieved with thrust from pectoral fins (or front limbs in marine mammals). Some fish propel themselves without a wave motion of the body, as in the slow-moving seahorses and *Gymnotus*. Other animals, such as cephalopods, use jet propulsion to travel fast, taking in water then squirting it back out in an explosive burst. Other swimming animals may rely predominantly on their limbs, much as humans do when swimming. Though life on land originated from the seas, terrestrial animals have returned to an aquatic lifestyle on several occasions, such as the fully aquatic cetaceans, now very distinct from their terrestrial ancestors.

Dolphins sometimes ride on the bow waves created by boats or surf on naturally breaking waves.

Benthic

Benthic locomotion is movement by animals that live on, in, or near the bottom of aquatic environments. In the sea, many animals walk over the seabed. Echinoderms primarily use their tube feet to move about. The tube feet typically have a tip shaped like a suction pad that can create a vacuum through contraction of muscles. This, along with some stickiness from the secretion of mucus, provides adhesion. Waves of tube feet contractions and relaxations move along the adherent surface and the animal moves slowly along. Some sea urchins also use their spines for benthic locomotion.

Crabs typically walk sideways (a behaviour that gives us the word *crabwise*). This is because of the articulation of the legs, which makes a sidelong gait more efficient. However, some crabs walk forwards or backwards, including raninids, *Libinia emarginata* and *Mictyris platycheles*. Some crabs, notably the Portunidae and Matutidae, are also capable of swimming, the Portunidae especially so as their last pair of walking legs are flattened into swimming paddles.

A stomatopod, *Nannosquilla decemspinosa*, can escape by rolling itself into a self-propelled wheel and somersault backwards at a speed of 72 rpm. They can travel more than 2 m using this unusual method of locomotion.

Aquatic Surface

Velella moves by sailing.

Velella, the by-the-wind sailor, is a cnidarian with no means of propulsion other than sailing. A small rigid sail projects into the air and catches the wind. *Velella* sails always align along the direction of the wind where the sail may act as an aerofoil, so that the animals tend to sail downwind at a small angle to the wind.

While larger animals such as ducks can move on water by floating, some small animals move across it without breaking through the surface. This surface locomotion takes advantage of the surface tension of water. Animals that move in such a way include the water strider. Water striders have legs that are hydrophobic, preventing them from interfering with the structure of water. Another form of locomotion (in which the surface layer is broken) is used by the basilisk lizard.

Aerial

Active Flight

A pair of brimstone butterflies in flight. The female, above, is in fast forward flight with a small angle of attack; the male, below, is twisting his wings sharply upward to gain lift and fly up towards the female.

Gravity is the primary obstacle to flight. Because it is impossible for any organism to have a density as low as that of air, flying animals must generate enough lift to ascend and remain airborne. One way to achieve this is with wings, which when moved through the air generate an upward lift force on the animal's body. Flying animals must be very light to achieve flight, the largest living flying animals being birds of around 20 kilograms. Other structural adaptations of flying animals include reduced and redistributed body weight, fusiform shape and powerful flight muscles; there may also be physiological adaptations. Active flight has independently evolved at least four times, in the insects, pterosaurs, birds, and bats. Insects were the first taxon to evolve flight, approximately 400 million years ago (mya), followed by pterosaurs approximately 220 mya, birds approximately 160 mya, then bats about 60 mya.

Gliding

Rather than active flight, some (semi-) arboreal animals reduce their rate of falling by gliding. Gliding is heavier-than-air flight without the use of thrust; the term "volplaning" also refers to this mode of flight in animals. This mode of flight involves flying a greater distance horizontally than vertically and therefore can be distinguished from a simple descent like a parachute. Gliding has evolved on more occasions than active flight. There are examples of gliding animals in several major taxonomic classes such as the invertebrates (e.g., gliding ants), reptiles (e.g., banded flying snake), amphibians (e.g., flying frog), mammals (e.g., sugar glider, squirrel glider).

Flying fish taking off.

Some aquatic animals also regularly use gliding, for example, flying fish, octopus and squid. The flights of flying fish are typically around 50 meters (160 ft), though they can use updrafts at the leading edge of waves to cover distances of up to 400 m (1,300 ft). To glide upward out of the water, a flying fish moves its tail up to 70 times per second. Several oceanic squid, such as the Pacific flying squid, leap out of the water to escape predators, an adaptation similar to that of flying fish. Smaller squids fly in shoals, and have been observed to cover distances as long as 50 m. Small fins towards the back of the mantle help stabilize the motion of flight. They exit the water by expelling water out of their funnel, indeed some squid have been observed to continue jetting water while airborne providing thrust even after leaving the water. This may make flying squid the only animals with jet-propelled aerial locomotion. The neon flying squid has been observed to glide for distances over 30 m, at speeds of up to 11.2 m/s.

Soaring

Soaring birds can maintain flight without wing flapping, using rising air currents. Many gliding birds are able to "lock" their extended wings by means of a specialized tendon. Soaring birds may alternate glides with periods of soaring in rising air. Five principal types of lift are used: thermals, ridge lift, lee waves, convergences and dynamic soaring.

Examples of soaring flight by birds are the use of:

- Thermals and convergences by raptors such as vultures.
- Ridge lift by gulls near cliffs.
- Wave lift by migrating birds.
- Dynamic effects near the surface of the sea by albatrosses.

Ballooning

Ballooning is a method of locomotion used by spiders. Certain silk-producing arthropods, mostly small or young spiders, secrete a special light-weight gossamer silk for ballooning, sometimes traveling great distances at high altitude.

Terrestrial

Pacific leaping blenny *Alticus arnoldorum* hopping.

Pacific leaping blenny climbing up a vertical piece of Plexiglas.

Forms of locomotion on land include walking, running, hopping or jumping, dragging and crawling or slithering. Here friction and buoyancy are no longer an issue, but a strong skeletal and muscular framework are required in most terrestrial animals for structural support. Each step also requires much energy to overcome inertia, and animals can store elastic potential energy in their tendons to help overcome this. Balance is also required for movement on land. Human infants learn to crawl first before they are able to stand on two feet, which requires good coordination as well as physical development. Humans are bipedal animals, standing on two feet and keeping one on the ground at all times while walking. When running, only one foot is on the ground at any one time at most, and both leave the ground briefly. At higher speeds momentum helps keep the body upright, so more energy can be used in movement.

Jumping

Jumping (saltation) can be distinguished from running, galloping, and other gaits where the entire body is temporarily airborne by the relatively long duration of the aerial phase and high angle of initial launch. Many terrestrial animals use jumping (including hopping or leaping) to escape predators or catch prey—however, relatively few animals use this as a primary mode of locomotion. Those that do include the kangaroo and other macropods, rabbit, hare, jerboa, hopping mouse, and kangaroo rat. Kangaroo rats often leap 2 m and reportedly up to 2.75 m at speeds up to almost 3 m/s (6.7 mph). They can quickly change their direction between jumps. The rapid locomotion of the banner-tailed kangaroo rat may minimize energy cost and predation risk. Its use of a "move-freeze" mode may also make it less conspicuous to nocturnal predators. Frogs are, relative to their size, the best jumpers of all vertebrates. The Australian rocket frog, *Litoria nasuta*, can leap over 2 metres (6 ft 7 in), more than fifty times its body length.

Peristalsis

Other animals move in terrestrial habitats without the aid of legs. Earthworms crawl by a peristalsis, the same rhythmic contractions that propel food through the digestive tract.

Sliding

Due to its low coefficient of friction, ice provides the opportunity for other modes of locomotion. Penguins either waddle on their feet or slide on their bellies across the snow, a movement called *tobogganing*, which conserves energy while moving quickly. Some pinnipeds perform a similar behaviour called *sledding*.

Brachiation

Some animals are specialized for moving on non-horizontal surfaces. One common habitat for such climbing animals is in trees, for example the gibbon is specialized for arboreal movement, traveling rapidly by brachiation. Another case is animals like the snow leopard living on steep rock faces such as are found in mountains. Some light animals are able to climb up smooth sheer surfaces or hang upside down by adhesion. Many insects can do this, though much larger animals such as geckos can also perform similar feats.

Walking and Running

Species have different numbers of legs resulting in large differences in locomotion.

Modern birds, though classified as tetrapods, usually have only two functional legs, which some (e.g., ostrich, emu, kiwi) use as their primary, Bipedal, mode of locomotion. Few modern mammalian species are habitual bipeds whose normal method of locomotion is two-legged. These include the macropods, kangaroo rats and mice, springhare, hopping mice, pangolins and homininan apes. Bipedalism is rarely found outside terrestrial animals—though at least two types of octopus walk bipedally on the sea floor using two of their arms, so they can use the remaining arms to camouflage themselves as a mat of algae or floating coconut.

Bipedal ostrich.

There are no three-legged animals—though some macropods, such as kangaroos, that alternate between resting their weight on their muscular tails and their two hind legs, could be looked at as an example of tripedal locomotion in animals.

Many familiar animals are quadrupedal, walking or running on four legs. A few birds use quadrupedal movement in some circumstances. For example, the shoebill sometimes uses its wings to right itself after lunging at prey. The newly hatched hoatzin bird has claws on its thumb and first finger enabling it to dexterously climb tree branches until its wings are strong enough for sustained flight. These claws are gone by the time the bird reaches adulthood.

Hexapedal stick-insect.

A relatively few animals use five limbs for locomotion. Prehensile quadrupeds may use their tail to assist in locomotion and when grazing, the kangaroos and other macropods use their tail to propel themselves forward with the four legs used to maintain balance.

Insects generally walk with six legs—though some insects such as nymphalid butterflies do not use the front legs for walking.

Arachnids have eight legs. Most arachnids lack extensor muscles in the distal joints of their appendages. Spiders and whipscorpions extend their limbs hydraulically using the pressure of their hemolymph. Solifuges and some harvestmen extend their knees by the use of highly elastic thickenings in the joint cuticle. Scorpions, pseudoscorpions and some harvestmen have evolved muscles that extend two leg joints (the femur-patella and patella-tibia joints) at once.

Octopedal locomotion by a spider.

The scorpion *Hadrurus arizonensis* walks by using two groups of legs (left 1, right 2, Left 3, Right 4 and Right 1, Left 2, Right 3, Left 4) in a reciprocating fashion. This alternating tetrapod coordination is used over all walking speeds.

Centipedes and millipedes have many sets of legs that move in metachronal rhythm. Some echinoderms locomote using the many tube feet on the underside of their arms. Although the tube feet resemble suction cups in appearance, the gripping action is a function of adhesive chemicals rather than suction. Other chemicals and relaxation of the ampullae allow for release from the substrate. The tube feet latch on to surfaces and move in a wave, with one arm section attaching to the surface as another releases. Some multi-armed, fast-moving starfish such as the sunflower seastar

(*Pycnopodia helianthoides*) pull themselves along with some of their arms while letting others trail behind. Other starfish turn up the tips of their arms while moving, which exposes the sensory tube feet and eyespot to external stimuli. Most starfish cannot move quickly, a typical speed being that of the leather star (*Dermasterias imbricata*), which can manage just 15 cm (6 in) in a minute. Some burrowing species from the genera *Astropecten* and *Luidia* have points rather than suckers on their long tube feet and are capable of much more rapid motion, "gliding" across the ocean floor. The sand star (*Luidia foliolata*) can travel at a speed of 2.8 m (9 ft 2 in) per minute. Sunflower starfish are quick, efficient hunters, moving at a speed of 1 m/min (3.3 ft/min) using 15,000 tube feet.

Many animals temporarily change the number of legs they use for locomotion in different circumstances. For example, many quadrupedal animals switch to bipedalism to reach low-level browse on trees. The genus of *Basiliscus* are arboreal lizards that usually use quadrupedalism in the trees. When frightened, they can drop to water below and run across the surface on their hind limbs at about 1.5 m/s for a distance of approximately 4.5 metres (15 ft) before they sink to all fours and swim. They can also sustain themselves on all fours while "water-walking" to increase the distance travelled above the surface by about 1.3 metres. When cockroaches run rapidly, they rear up on their two hind legs like bipedal humans; this allows them to run at speeds up to 50 body lengths/s, equivalent to a "couple hundred miles per hour, if you scale up to the size of humans". When grazing, kangaroos use a form of pentapedalism (four legs plus the tail) but switch to hopping (bipedalism) when they wish to move at a greater speed.

Multi-legged millipede.

Powered Cartwheeling

The Moroccan flic-flac spider (*Cebrennus rechenbergi*) uses a series of rapid, acrobatic flic-flac movements of its legs similar to those used by gymnasts, to actively propel itself off the ground, allowing it to move both down and uphill, even at a 40 percent incline. This behaviour is different than other huntsman spiders, such as *Carparachne aureoflava* from the Namib Desert, which uses passive cartwheeling as a form of locomotion. The flic-flac spider can reach speeds of up to 2 m/s using forward or back flips to evade threats.

Subterranean

Some animals move through solids such as soil by burrowing using peristalsis, as in earthworms, or other methods. In loose solids such as sand some animals, such as the golden mole, marsupial mole, and the pink fairy armadillo, are able to move more rapidly, 'swimming' through the loose substrate. Burrowing animals include moles, ground squirrels, naked mole-rats, tilefish, and mole crickets.

Arboreal Locomotion

A brachiating gibbon.

Arboreal locomotion is the locomotion of animals in trees. Some animals may only scale trees occasionally, while others are exclusively arboreal. These habitats pose numerous mechanical challenges to animals moving through them, leading to a variety of anatomical, behavioural and ecological consequences as well as variations throughout different species. Furthermore, many of these same principles may be applied to climbing without trees, such as on rock piles or mountains. The earliest known tetrapod with specializations that adapted it for climbing trees, was *Suminia*, a synapsid of the late Permian, about 260 million years ago. Some invertebrate animals are exclusively arboreal in habitat, for example, the tree snail.

Brachiation is a form of arboreal locomotion in which primates swing from tree limb to tree limb using only their arms. During brachiation, the body is alternately supported under each forelimb. This is the primary means of locomotion for the small gibbons and siamangs of southeast Asia. Some New World monkeys such as spider monkeys and muriquis are "semibrachiators" and move through the trees with a combination of leaping and brachiation. Some New World species also practice suspensory behaviors by using their prehensile tail, which acts as a fifth grasping hand.

Energetics

Animal locomotion requires energy to overcome various forces including friction, drag, inertia and gravity, although the influence of these depends on the circumstances. In terrestrial environments, gravity must be overcome whereas the drag of air has little influence. In aqueous environments, friction (or drag) becomes the major energetic challenge with gravity being less of an influence. Remaining in the aqueous environment, animals with natural buoyancy expend little energy to maintain a vertical position in a water column. Others naturally sink, and must spend energy to remain afloat. Drag is also an energetic influence in flight, and the aerodynamically efficient body shapes of flying birds indicate how they have evolved to cope with this. Limbless organisms moving on land must energetically overcome surface friction, however, they do not usually need to expend significant energy to counteract gravity.

Newton's third law of motion is widely used in the study of animal locomotion: if at rest, to move forwards an animal must push something backwards. Terrestrial animals must push the solid ground, swimming and flying animals must push against a fluid (either water or air). The effect of forces during locomotion on the design of the skeletal system is also important, as is the interaction between locomotion and muscle physiology, in determining how the structures and effectors of locomotion enable or limit animal movement. The energetics of locomotion involves the energy expenditure by animals in moving. Energy consumed in locomotion is not available for other efforts, so animals typically have evolved to use the minimum energy possible during movement. However, in the case of certain behaviors, such as locomotion to escape a predator, performance (such as speed or maneuverability) is more crucial, and such movements may be energetically expensive. Furthermore, animals may use energetically expensive methods of locomotion when environmental conditions (such as being within a burrow) preclude other modes.

The most common metric of energy use during locomotion is the net [also termed "incremental"] cost of transport, defined as the amount of energy (e.g., Joules) needed above baseline metabolic rate to move a given distance. For aerobic locomotion, most animals have a nearly constant cost of transport - moving a given distance requires the same caloric expenditure, regardless of speed. This constancy is usually accomplished by changes in gait. The net cost of transport of swimming is lowest, followed by flight, with terrestrial limbed locomotion being the most expensive per unit distance. However, because of the speeds involved, flight requires the most energy per unit time. This does not mean that an animal that normally moves by running would be a more efficient swimmer; however, these comparisons assume an animal is specialized for that form of motion. Another consideration here is body mass—heavier animals, though using more total energy, require less energy *per unit mass* to move. Physiologists generally measure energy use by the amount of oxygen consumed, or the amount of carbon dioxide produced, in an animal's respiration. In terrestrial animals, the cost of transport is typically measured while they walk or run on a motorized treadmill, either wearing a mask to capture gas exchange or with the entire treadmill enclosed in a metabolic chamber. For small rodents, such as deer mice, the cost of transport has also been measured during voluntary wheel running.

Energetics is important for explaining the evolution of foraging economic decisions in organisms; for example, a study of the African honey bee, *A. m. scutellata*, has shown that honey bees may trade-off the high sucrose content of viscous nectar for the energetic benefits of warmer, less concentrated nectar, which also reduces their consumption and flight time.

Passive Locomotion

Passive locomotion in animals is a type of mobility in which the animal depends on their environment for transportation.

Hydrozoans

The Portuguese man o' war (*Physalia physalis*) lives at the surface of the ocean. The gas-filled bladder, or pneumatophore (sometimes called a "sail"), remains at the surface, while the remainder is submerged. Because the Portuguese man o' war has no means of propulsion, it is moved by a combination of winds, currents, and tides. The sail is equipped with a siphon. In the event of a surface attack, the sail can be deflated, allowing the organism to briefly submerge.

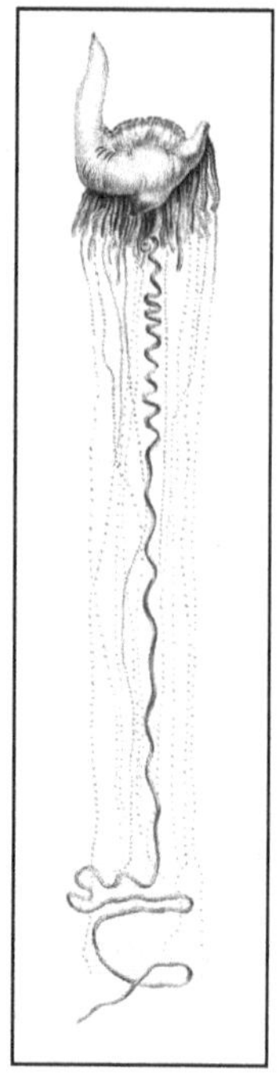

Physalia physalis

Arachnids

The wheel spider (*Carparachne aureoflava*) is a huntsman spider approximately 20 mm in size and native to the Namib Desert of Southern Africa. The spider escapes parasitic pompilid wasps by flipping onto its side and cartwheeling down sand dunes at speeds of up to 44 turns per second. If the spider is on a sloped dune, its rolling speed may be 1 metre per second.

A spider (usually limited to individuals of a small species), or spiderling after hatching, climbs as high as it can, stands on raised legs with its abdomen pointed upwards ("tiptoeing"), and then releases several silk threads from its spinnerets into the air. These form a triangle-shaped parachute that carries the spider on updrafts of winds, where even the slightest breeze transports it. The Earth's static electric field may also provide lift in windless conditions.

Insects

The larva of *Cicindela dorsalis*, the eastern beach tiger beetle, is notable for its ability to leap into the air, loop its body into a rotating wheel and roll along the sand at a high speed using wind to propel itself. If the wind is strong enough, the larva can cover up to 60 metres (200 ft) in this manner. This remarkable ability may have evolved to help the larva escape predators such as the tiphiid wasp *Methocha*.

Members of the largest subfamily of cuckoo wasps, Chrysidinae, are generally kleptoparasites, laying their eggs in host nests, where their larvae consume the host egg or larva while it is still young. Chrysidines are distinguished from the members of other subfamilies in that most have flattened or concave lower abdomens and can curl into a defensive ball when attacked by a potential host, a process known as conglobation. Protected by hard chitin in this position, they are expelled from the nest without injury and can search for a less hostile host.

Fleas can jump vertically up to 18 cm and horizontally up to 33 cm, however, although this form of locomotion is initiated by the flea, it has little control of the jump - they always jump in the same direction, with very little variation in the trajectory between individual jumps.

Crustaceans

Although stomatopods typically display the standard locomotion types as seen in true shrimp and lobsters, one species, *Nannosquilla decemspinosa*, has been observed flipping itself into a crude wheel. The species lives in shallow, sandy areas. At low tides, *N. decemspinosa* is often stranded by its short rear legs, which are sufficient for locomotion when the body is supported by water, but not on dry land. The mantis shrimp then performs a forward flip in an attempt to roll towards the next tide pool. *N. decemspinosa* has been observed to roll repeatedly for 2 metres (6.6 ft), but they typically travel less than 1 m (3.3 ft). Again, the animal initiates the movement but has little control during its locomotion.

Animal Transport

Some animals change location because they are attached to, or reside on, another animal or moving structure. This is arguably more accurately termed "animal transport".

Remoras

Some remoras, such as this Echeneis naucrates, may attach themselves to scuba divers.

Remoras are a family (*Echeneidae*) of ray-finned fish. They grow to 30–90 cm (0.98–2.95 ft) long, and their distinctive first dorsal fins take the form of a modified oval, sucker-like organ with slat-like structures that open and close to create suction and take a firm hold against the skin of larger marine animals. By sliding backward, the remora can increase the suction, or it can release itself by swimming forward. Remoras sometimes attach to small boats. They swim well on their own, with a sinuous, or curved, motion. When the remora reaches about 3 cm (1.2 in), the disc is fully formed and the remora can then attach to other animals. The remora's lower jaw projects beyond the upper, and the animal lacks a swim bladder. Some remoras associate primarily with specific host species. They are commonly found attached to sharks, manta rays, whales, turtles, and dugongs. Smaller remoras also fasten onto fish such as tuna and swordfish, and some small remoras travel in the mouths or gills of large manta rays, ocean sunfish, swordfish, and sailfish. The remora benefits by using the host as transport and protection, and also feeds on materials dropped by the host.

Angler Fish

In some species of anglerfish, when a male finds a female, he bites into her skin, and releases an enzyme that digests the skin of his mouth and her body, fusing the pair down to the blood-vessel level. The male becomes dependent on the female host for survival by receiving nutrients via their shared circulatory system, and provides sperm to the female in return. After fusing, males increase in volume and become much larger relative to free-living males of the species. They live and remain reproductively functional as long as the female lives, and can take part in multiple spawnings. This extreme sexual dimorphism ensures, when the female is ready to spawn, she has a mate immediately available. Multiple males can be incorporated into a single individual female with up to eight males in some species, though some taxa appear to have a one male per female rule.

Parasites

Many endoparasites and ectoparasites, due to their parasitic behaviour, are transported by other animals. For example, tapeworms attach themselves to the inside of the alimentary tracts of other animals and do not locomote within the animal. They do however depend on movement of the host to distribute their eggs.

Other parasites may locomote within, or on, their host, which in turn might be active or stationary. For example, an adult dog flea may crawl about the skin of its sleeping canine host (locomotion), but when the dog awakes and moves, it could be argued the flea is being transported.

Changes between Media

Some animals locomote between different media, e.g., from aquatic to arial. This often requires different modes of locomotion in the different media and may require a distinct transitional locomotor behaviour.

There are a large number of semi-aquatic animals (animals that spend part of their life cycle in water, or generally have part of their anatomy underwater). These represent the major taxons of mammals (e.g., beaver, otter, polar bear), birds (e.g., penguins, ducks), reptiles (e.g., anaconda, bog turtle, marine iguana) and amphibians (e.g., salamanders, frogs, newts).

Fish

Some fish use multiple modes of locomotion. Walking fish may swim freely or at other times "walk" along the ocean or river floor, but not on land (e.g., the flying gurnard —which does not actually fly—and batfishes of the Ogcocephalidae family). Amphibious fish, are fish that are able to leave water for extended periods of time. These fish use a range of terrestrial locomotory modes, such as lateral undulation, tripod-like walking (using paired fins and tail), and jumping. Many of these locomotory modes incorporate multiple combinations of pectoral, pelvic and tail fin movement. Examples include eels, mudskippers and the walking catfish. Flying fish can make powerful, self-propelled leaps out of water into air, where their long, wing-like fins enable gliding flight for considerable distances above the water's surface. This uncommon ability is a natural defense mechanism to evade predators. The flights of flying fish are typically around 50 meters, though they can use updrafts at the leading edge of waves to cover distances of up to 400 m (1,300 ft). They can travel at speeds of more than 70 km/h (43 mph). Maximum altitude is 6 m (20 ft) above the surface of the sea. Some accounts have them landing on ships' decks.

Marine Mammals

Pacific white-sided dolphins porpoising.

When swimming, several marine mammals such as dolphins, porpoises and pinnipeds, frequently leap above the water surface whilst maintaining horizontal locomotion. This is done for various reasons. When travelling, jumping can save dolphins and porpoises energy as there is less friction while in the air. This type of travel is known as "porpoising". Other reasons for dolphins and porpoises performing porpoising include orientation, social displays, fighting, non-verbal communication, entertainment and attempting to dislodge parasites. In pinnipeds, two types of porpoising have been identified. "High porpoising" is most often near (within 100 metres) the shore and is often followed by minor course changes; this may help seals get their bearings on beaching or rafting sites. "Low porpoising" is typically observed relatively far (more than 100 metres) from shore and often aborted in favour of anti-predator movements; this may be a way for seals to maximize sub-surface vigilance and thereby reduce their vulnerability to sharks.

Some whales raise their (entire) body vertically out of the water in a behaviour known as "breaching".

Birds

Some semi-aquatic birds use terrestrial locomotion, surface swimming, underwater swimming and flying (e.g., ducks, swans). Diving birds also use diving locomotion (e.g., dippers, aulks). Some birds (e.g., ratites) have lost the primary locomotion of flight. The largest of these, ostriches, when being pursued by a predator, have been known to reach speeds over 70 km/h (43 mph), and can maintain a steady speed of 50 km/h (31 mph), which makes the ostrich the world's fastest two-legged animal: Ostriches can also locomote by swimming. Penguins either waddle on their feet or slide on their bellies across the snow, a movement called *tobogganing*, which conserves energy while moving quickly. They also jump with both feet together if they want to move more quickly or cross steep or rocky terrain. To get onto land, penguins sometimes propel themselves upwards at a great speed to leap out the water.

Changes During the Life-cycle

An animal's mode of locomotion may change considerably during its life-cycle. Barnacles are exclusively marine and tend to live in shallow and tidal waters. They have two nektonic (active swimming) larval stages, but as adults, they are sessile (non-motile) suspension feeders. Frequently, adults are found attached to moving objects such as whales and ships, and are thereby transported (passive locomotion) around the oceans.

Function

Animals locomote for a variety of reasons, such as to find food, a mate, a suitable microhabitat, or to escape predators.

Food Procurement

Paddlefish ram suspension-feeding zooplankton in aquarium.

Animals use locomotion in a wide variety of ways to procure food. Terrestrial methods include ambush predation, social predation, grazing. Aquatic methods include filterfeeding, grazing, ram feeding, suction feeding, protrusion and pivot feeding. Other methods include parasitism and parasitoidism.

Methods of Study

Documentary film, shot at 1200 fps, used to study the locomoton of a cheetah. The end of the video shows the methods used for filming.

A variety of methods and equipment are used to study animal locomotion:

- Treadmills are used to allow animals to walk or run while remaining stationary with respect to external observers. This technique facilitates filming or recordings of physiological information from the animal (e.g., during studies of energetics). Motorized treadmills are also used to measure the endurance capacity (stamina) of animals.
- Racetracks lined with photocells or filmed while animals run along them are used to measure acceleration and maximal sprint speed.
- Kinematics is the study of the motion of an entire animal or parts of its body. It is typically accomplished by placing visual markers at particular anatomical locations on the animal and then recording video of its movement. The video is often captured from multiple angles, with frame rates exceeding 2000 frames per second when capturing high speed movement. The location of each marker is determined for each video frame, and data from multiple views is integrated to give positions of each point through time. Computers are sometimes used to track the markers, although this task must often be performed manually. The kinematic data can be used to determine fundamental motion attributes such as velocity, acceleration, joint angles, and the sequencing and timing of kinematic events. These fundamental attributes can be used to quantify various higher level attributes, such as the physical abilities of the animal (e.g., its maximum running speed, how steep a slope it can climb), neural control of locomotion, gait, and responses to environmental variation. These, in turn, can aid in formulation of hypotheses about the animal or locomotion in general.
- Force plates are platforms, usually part of a trackway, that can be used to measure the magnitude and direction of forces of an animal's step. When used with kinematics and a sufficiently detailed model of anatomy, inverse dynamics solutions can determine the forces not just at the contact with the ground, but at each joint in the limb.
- Electromyography (EMG) is a method of detecting the electrical activity that occurs when muscles are activated, thus determining which muscles an animal uses for a given movement. This can be accomplished either by surface electrodes (usually in large animals) or implanted electrodes (often wires thinner than a human hair). Furthermore, the intensity of electrical activity can correlate to the level of muscle activity, with greater activity implying (though not definitively showing) greater force.
- Sonomicrometry employs a pair of piezoelectric crystals implanted in a muscle or tendon to continuously measure the length of a muscle or tendon. This is useful because surface kinematics may be inaccurate due to skin movement. Similarly, if an elastic tendon is in series with the muscle, the muscle length may not be accurately reflected by the joint angle.
- Tendon force buckles measure the force produced by a single muscle by measuring the strain of a tendon. After the experiment, the tendon's elastic modulus is determined and used to compute the exact force produced by the muscle. However, this can only be used on muscles with long tendons.
- Particle image velocimetry is used in aquatic and aerial systems to measure the flow of fluid around and past a moving aquatic organism, allowing fluid dynamics calculations to determine pressure gradients, speeds, etc.

- Fluoroscopy allows real-time X-ray video, for precise kinematics of moving bones. Markers opaque to X-rays can allow simultaneous tracking of muscle length.

These methods can be combined. For example, studies frequently combine EMG and kinematics to determine *motor pattern,* the series of electrical and kinematic events that produce a given movement.

Role of Skin in Locomotion

Role of skin in locomotion describes how the integumentary system is involved in locomotion. Typically the integumentary system can be thought of as skin, however the integumentary system also includes the segmented exoskeleton in arthropods and feathers of birds. The primary role of the integumentary system is to provide protection for the body. However, the structure of the skin has evolved to aid animals in their different modes of locomotion. Soft bodied animals such as starfish rely on the arrangement of the fibers in their tube feet for movement. Eels, snakes, and fish use their skin like an external tendon to generate the propulsive forces need for undulatory locomotion. Vertebrates that fly, glide, and parachute also have a characteristic fiber arrangements of their flight membranes that allows for the skin to maintain its structural integrity during the stress and strain experienced during flight.

Soft Bodied Locomotion in Invertebrates

The arms of octopus are muscular hydrostats.

The term "Soft Bodied" refers to animals which lack typical systems of skeletal support - included in these are most insect larvae and true worms. Animals that are soft bodied are constrained by the geometry and form of their bodies. However it is the geometry and form of their bodies that generate the forces they need to move. The structure of soft bodied skin can be characterized by a patterned fiber arrangement, which provides the shape and structure for a soft bodied animals. Internal to the patterned fiber layer is typically a liquid filled cavity, which is used to generate hydrostatic pressures for movement. Some animals that exhibit soft bodied locomotion include starfish, octopus, and flatworms.

Hydrostatic Skeleton

A hydrostatic skeleton uses hydrostatic pressure generated from muscle contraction against a liquid filled cavity. The liquid filled cavity is commonly referred to as the hydrostatic body. The liquid within the hydrostatic body acts as an incompressible fluid and the body wall of the hydrostatic body provides a passive elastic antagonist to muscle contraction, which in turn generates a force, which in turn creates movement. This structure plays a role in invertebrate support and locomotor systems and is used for the tube feet in starfish and body of worms . A specialized version of the hydrostatic skeleton is a called a muscular hydrostat, which consists of a tightly packed array of three-dimensional muscle fibers surrounding a hydrostatic body. Examples of muscular hydrostats include the arms of octopus and elephant trunks.

Fiber Arrangement

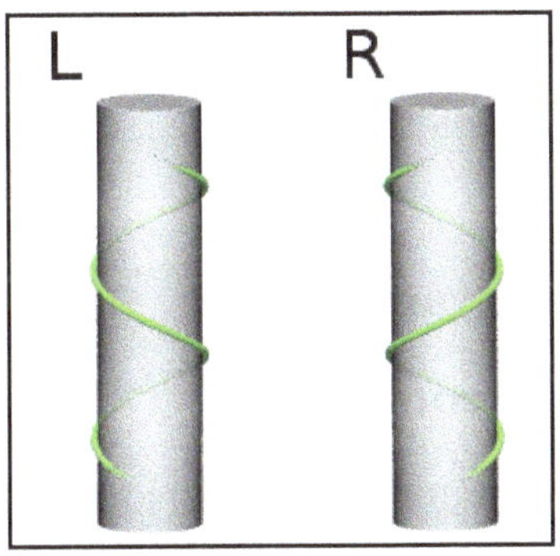

left- and right-handed helices.

The arrangement of the connective tissue fibers and muscle fibers create the skeletal support of a soft bodied animal. The arrangement of the fibers around a hydrostatic body limits the range of movement of the hydrostatic body (the "body" of a soft bodied animal) and defines the way the hydrostatic body moves.

Muscle Fibers

Typically muscle fibers surround the hydrostatic body. There are two main types of muscle fibers orientations that are responsible for the movement: the circular orientations and longitudinal orientations. Circular muscles decrease the diameter of a hydrostatic body, resulting in an increase in the length of the body, whereas longitudinal muscles shortens the length of a hydrostatic body, resulting in an increase in the diameter of the body. There are four categories of movements of a hydrostatic skeleton : elongation, shortening, bending and torsion. Elongation, which involves an increase in the length of a hydrostatic body requires either circular muscles, a transverse muscle arrangement, or radial muscle arrangement. For a transverse muscle arrangement, parallel sheets of muscle fibers that extend along the length of a hydrostatic body. For a radial muscle arrangement, radial muscles radiate from a central axis along the axis perpendicular to the long axis. Shortening involves the contraction of the longitudinal muscle. Both shortening and bending involve the contraction of longitudinal muscle, but for bending motion some of the antagonistic muscles work synergistically with longitudinal muscles. The amplitude of movements are based upon the antagonistic muscles forces and the amount of leverage the antagonistic muscle provides for movement. For the torsion motion, muscles are arranged in helical layers around a hydrostatic body. The fiber angle (the angle the fiber makes with the long axis of the body) plays a critical role in torsion, if the angle is greater than 54°44', during muscle contraction, torsion and elongation will occur. If the fiber angle is less than 54°44', torsion and shortening will occur.

Connective Tissue Fibers

The arrangement of connective tissue fibers determines the range of motion of a body, and serves as an antagonist against muscle contraction. The most commonly observed connective tissue arrangement for soft bodied animals consists of layers of alternating right and left-handed helices of connective tissue fibers which surround the hydraulic body. This cross helical arrangement is seen in the tube feet starfish, different types of worms and suckers in octopus. This cross helical arrangement allows for the connective tissue layers to evenly distribute force throughout the hydrostatic body. Another commonly observed connective tissue fiber range is when the connective tissue fibers are embedded within a muscle layer. This arrangement of connective tissue fibers creates a stiffer body wall and more muscle antagonism, which allows for more elastic force to be generated and released during movement. This fiber arrangement is seen in the mantle of squid and the fins in sharks.

Specialized Function in Vertebrates

Swimming and Undulatory Locomotion

The skin of these animal that use undulatory motion to locomote have several distinct characteristics. The skin of these animals consists of **cro**ss-helical arrangement of collagen and elastin fibers embedded in the dermal layer of skin, a two-dimensional stiffness. which permits bending at small curvatures and resists bending at high curvatures and skin is attached directly to the underlying muscles. Fish, shark, and snakes are all examples of animals that locomote using undulatory locomotion.

Eel

Closeup image of the skin of a marbled eel.

The cross helical fiber arrangement of the two dermal fibers types collagen and elastin, are responsible for the mechanical properties of the skin such as the two dimensional stiffness seen in the eel skin. In the longitudinal direction, eel skin behaves like a pure fiber system, with a lessertensile strength than skin in the hoop direction. The skin in the hoop direction exhibits a higher elastic modulus than the skin in the longitudinal direction. The two dimensional stiffness allows for the body of the eel to be modeled a pressurized cylinder with the fiber angle of the cross helical arrangement dictating the method by which the eel moves. Eel skin behaves like skin having a fiber angle greater than 45°. In an eel with the cross helical fiber arrangement, muscle contraction in the anterior region bends the fish, and so the skin on the convex side is extended in the longitudinal direction. The extension in the longitudinal direction produces contraction in the hoop direction as the fiber angle decreases until these dimensional changes are resisted by the body of the eel. The skin becomes skin, and additional longitudinal force(applied by skin) results in force being transmitted along the tail. Therefore, changes in fiber angle of the cross helical arrangement in eel skin allows for the transmission of force through the skin during swimming. The skin act like an external tendon allowing for an eel to generate a greater propulsive force per muscle contraction. In addition to the eel skin acting as an external tendon, the skin attaches directly to the underlying muscle, which allow for the eel to generate an even greater force per muscle contraction.

Longnose Gar

Closeup of the skin of the longnose gar.

Due to the heavily scaled skin of the Longnose gar, some of the mechanical properties differ from model of describing how eel skin adds movement. The scale row resists longitudinal forces, which unlike eel skin, makes the skin stiffer in the longitudinal direction, providing myomeres with leverage and anchorage for pulling tendons. At low curvatures, it appears that the dermis is slack on both the concave and convex sides of the body. When the dermis is placed in tension, and resistance to bending is developed, which is referred to as flexural stiffness of the fish skin. The flexural stiffness is a result of the two dimensional stiffness of fish with heavily scaled skin, such as the longnose gar. This mechanical property of fish skin is important to the way a fish swims, because this mechanical property passively stiffens the body, which would otherwise would have been done muscularly. The flexural stiffness of fish skin act in a manner similar to the mechanism by which eel skin acts as an external tendon, however in the case of fish skin, the flexural stiffness acts as a mechanism to decelerate body movement rather than to generate a propulsive force.

Snake

Snakes are one of the few vertebrates in which the skin alone is sufficient for locomotion. During Rectilinear locomotion, the skeleton remains fixed, while the skin is alternately lifted and pulled forward, and then allowed to contact the ground and pulled backwards, propelling the body forward.

One of the interesting aspects of snake skin are folds of intersquamous skin between longitudinally oriented scale rows. The function of these folds is to permit the circumference of the snake to increase, allowing prey to pass into the stomach during feeling. Snakes differ from eels in the direction in which the skin is stiffer, the dorsal scale rows are more flexible in snake than in eels because the dorsal scale row associated with stretching. Differences in the local dermal structures, such as variations in the diameters and orientation of collagen fibers within the intersquamous skin create local differences in the mechanical properties of the snake skin, thus allowing it to adapt to the stresses and strains during the feeding process.

Aerial Locomotion

Gliding, Flying and Parachuting are some of the some methods of aerial locomotion used by animals. Vertebrates have altered the structure of the skin to accommodate the stresses and strains of flight. Typically mammalian skin consists of collagen fibers arranged in a felt-work pattern, with no preferential fiber orientation. However,the structures of skin in bats, birds, and gliding lizards are very different from those of typical mammalian skin. The structural arrangement of the fibers within bat wing skin enables the bat to act like a spring during the down-stroke of flapping. The scales of gliding lizards are arranged in a regular rib like pattern to enable to lizard to act as an airfoil. Avain skin must be structurally arranged such that "the coat of feathers" remains smooth and intact during flight.

Bats

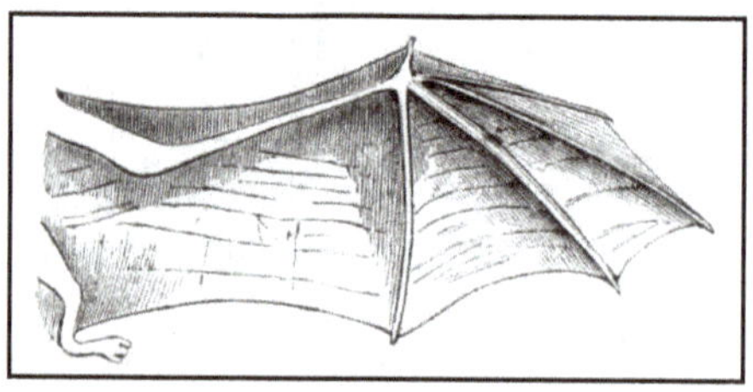

Drawing of the bat wing skin showing the fibers of "mesh like scaffolding".

A closeup view of the felt like fiber pattern seen in elephant skin.

Bats rely on skin on their wings to generate lift and thrust used in flight. Therefore, the structure of the bat wing skin is different from the skin of the bat body. Bat wing skin consists of two thin layers of epidermis with a thin layer of dermis/hypodermis located between the epidermal layers whereas the skin of the bat body consists of a single layer of epidermis with a thicker layer of dermis internal to the epidermis. Within the dermal and epidermal layer of bat wing skin, the connective tissue and muscle fibers provide the structural support. The connective tissue fibers within bat wing skin consists of collagen and elastin fiber bundles arranged in a "regular mesh like scaffolding", which the nerves, skeletal muscle fibers and blood vessels embed themselves into. Of the muscles that insert themselves into the mesh scaffolding, larger muscles anchor the skin to the bone and control the membrane tension and camber of the bat wing during flight, whereas smaller muscles, which originate from within the mesh scaffolding, attach to collagen fibers within the fiber network and modulate bone loading and allow for precise control of wing shape and tension. As seen in snakes, local structural differences within the arrangement of the fibers change the mechanical properties of local area, but there are general characteristics that describe the mechanical behavior of bat wing skin. Within the mesh scaffolding of bat wing skin, collagen fibers cross bones perpendicular to the long axes of the bones, therefore mechanical properties of bat wing skin oriented perpendicular to the long axes of the bones exhibit a lower stiffness than the skin that is oriented parallel to the long axes of the bodies. Stiffer skin is necessary for bat wing skin oriented in the direction parallel to the long axes of the bones to prevent too much deformation of bat wing skin during flight(with respect to the bone), resulting in the shearing of the bat wing skin off of the bone. Flexible skin is necessary for the direction perpendicular to the long axes of the bones for facilitating the shape changes needed for movement and control during flight. This anisotropy of bat wing skin is also useful as method of storing and releasing elastic energy, particularly during the downstroke. During the downstroke, the bat extends its wing and the wing skin experiences an aerodynamic force. The wing skin expands and counteracts the aerodynamic force. After the wing is fully extended, the orientation of the wing is nearly vertical and the aerodynamic forces on the wing are reduced. As the aerodynamic force is reduced, the wing recoils, drawing the digits of the wings together in preparation for the upstroke.

Gliding Lizards

There are two different mechanisms by which lizards glide through the air. Both mechanisms involve the patagia.

Active parachuting mechanism.

In the active mechanism, skeletal supports and muscles run through the patagia of lizards. The skeletal supports and muscle erect the flight membrane and control the gliding using the patagia. Most of the lizards that exhibit this active gliding mechanism are agamine lizards such lizards in the genus Draco. For the passive mechanism of gliding in lizards, the patagia is unfurled by air pressure alone. The patagia of the passive mechanism differs from patagia of the active mechanism; there is the lack of skeleton support and musculature in patagia of the gliding lizards with the passive gliding mechanism . The passive mechanism of gliding is seen in smaller lizards such as the geckos of the genus Ptychozoon. For the passive mechanism of gliding, body movements are believed to control the descent of the gliding lizard.

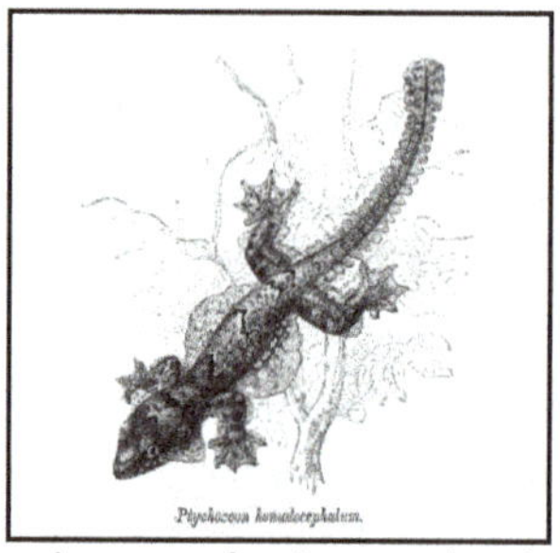

Passive parachuting mechanism.

The surface area to body ratios of lizards with different gliding mechanisms are similar, but how the surface area is distributed is different. The difference in the distribution of surface area indicates the differences in the role of the patagia and accessory areas for the different flight mechanisms. Lizards with passive gliding mechanisms tend to have smaller patagia relative to lizards with active flight mechanisms. However,lizards with passive flight mechanism have,ore surface area located in accessory areas(i.e. webbed toes,tail) than lizards with the active gliding mechanism.

The structure of the skin of the patagia and accessory areas for the patagia of a Ptychozoon kuhli, flying gecko, exhibiting the passive gliding mechanism consists of five layers; a layer of adipose tissue is surrounded by a layer of dermis on each side (ventral and dorsal) and a layer of epidermis is external to the two dermal layers. The distribution of the adipose tissue IS thickest close to the body wall. This thick layer of adipose tissue at close to the body wall is believed to provide a "safety factor" for the structural elements of the skin (i.e. collagen fibers) near the body wall. The thick layer of adipose tissue is more compliant than the structural elements of the body wall (i.e. ribs, muscles), therefore will more readily deform(absorb force) before the structural elements of the skin experience a force. The layer of adipose tissue also aids in the creation of the domed and cambered shape of the patagia. With regards to the structure of the dermal layer of the patagia, there is a thick layer of collagen fibers oriented parallel to the axis of the patagial ribs. These collagen fibers act as the structural support for the shape of the patagia, and provide the stiffness necessary to resist shape chang . The most prominent features of the epidermal layer of the patagia are the scales. The morphology of the dorsal scales of the patagia change as a result of their functional role. A large portion of the dorsal scales of the patagia are arranged in regular rib-like pattern, which guide the flow of air and allow for the lizard to behave as an airfoil. However at the hinge joints (places where patagia folds and attaches to limbs), the regular rib like structure of scales breakdown into a more random distribution of scales. This breakdown of scales is believed to aid in the mechanical loading of the patagia during the unfurling process and also determining the extent the patagia unfurling during flight.

Birds

The wing of a woodpecker.

Avian skin is a bit more complicated than the skin of gliding lizards or bats because the presence of feathers. In addition to the counteracting stresses and stains associated with flight, avian skin must provide a means to monitor and anchor a "coat of feathers", thus the structure of avian skin is different from skin of other flying and gliding animals. To better understand the structure of avian skin, avian skin has been broken down into three different functional components:

- Hydraulic skeleto-muscular apparatus of the feathers

This functional component consists of the only of the structural features of the feather tracts, namely the cutis, and connective tissue layer fascia superficialis. This functional component was named "hydraulic skeletal" due to the fact that the fat bodies embedded within cutis and fascia act similar to the hydrostatic bodies within a hydrostatic skeleton. However the functional role of the fat bodies within the hydraulic skeleto-muscular apparatus of the feathers is to counteract forces generated by the erector and depressor muscle of the feathers tracts. rather than to facilitate movement within of a body.

- Dermo-subcutaneous muscular system of integument

This functional component of avian skin consists of the smooth muscle of the apertia and striated subcutaneous muscles. The smooth muscles of the apertia counteract the horizontal forces experienced by the feather follicles. The striated subcutaneous muscles also adjust the position of the feather follicles in the directions the smooth muscle cannot. Together this system acts as an integrated muscular system that properly positions the feather tracts on the body of the bird.

- Subcutaneous hydraulic skeletal system

This functional component of avian skin consists of the fat bodies of the fascia superficialis and Fascia subcutanea. The majority of the fat bodies are located either between fascia superficialis and the Fascia subcutanea. These fat bodies are stratically located at depression within the body of the bird and function to even out depressions so that feather tracts of the skeleto-muscular apparatus function properly.

Arboreal Locomotion

Arboreal locomotion is the locomotion of animals in trees. In habitats in which trees are present, animals have evolved to move in them. Some animals may scale trees only occasionally, but others are exclusively arboreal. The habitats pose numerous mechanical challenges to animals moving through them and lead to a variety of anatomical, behavioral and ecological consequences as well as variations throughout different species. Furthermore, many of these same principles may be applied to climbing without trees, such as on rock piles or mountains.

The earliest known tetrapod with specializations that adapted it for climbing trees was *Suminia*, a synapsid of the late Permian, about 260 million years ago.

Some invertebrate animals are exclusively arboreal in habitat, such as the tree snail.

Leopards are good climbers and can carry their kills up their trees to keep them out of reach from scavengers and other predators.

Biomechanics

Arboreal habitats pose numerous mechanical challenges to animals moving in them, which have been solved in diverse ways. These challenges include moving on narrow branches, moving up and down inclines, balancing, crossing gaps, and dealing with obstructions.

Diameter

Moving along a narrow surface poses special difficulties to animals. During locomotion on the ground, the location of the center of mass may swing from side to side, but during arboreal locomotion, this would result in the center of mass moving beyond the edge of the branch, resulting in a tendency to topple over. Additionally, foot placement is constrained by the need to make contact with the narrow branch. This narrowness severely restricts the range of movements and postures an animal can use to move.

Incline

Branches are frequently oriented at an angle to gravity in arboreal habitats, including being vertical, which poses special problems. As an animal moves up an inclined branch, they must fight the force of gravity to raise their body, making movement more difficult. Conversely, as the animal descends, it must also fight gravity to control its descent and prevent falling. Descent can be particularly problematic for many animals, and highly arboreal species often have specialized methods for controlling their descent.

Balance

Due to the height of many branches and the potentially disastrous consequences of a fall, balance is of primary importance to arboreal animals. On horizontal and gently sloped branches, the primary problem is tipping to the side due to the narrow base of support. The narrower the branch, the greater the difficulty in balancing a given animal faces. On steep and vertical branches, tipping becomes less of an issue, and pitching backwards or slipping downwards becomes the most likely failure. In this case, large-diameter branches pose a greater challenge, since the animal cannot place its forelimbs closer to the center of the branch than its hindlimbs.

Gibbons are very good brachiators because their elongated limbs enable them to easily swing and grasp on to branches.

Crossing Gaps

Branches are not continuous, and any arboreal animal must be able to move between gaps in the branches, or even between trees. This can be accomplished by reaching across gaps, by leaping across them or gliding between them.

Obstructions

Arboreal habitats often contain many obstructions, both in the form of branches emerging from the one being moved on and other branches impinging on the space the animal needs to move through. These obstructions may impede locomotion, or may be used as additional contact points to enhance it. While obstructions tend to impede limbed animals, they benefit snakes by providing anchor points.

Anatomical Specializations

Arboreal organisms display many specializations for dealing with the mechanical challenges of moving through their habitats.

Limb Length

Arboreal animals frequently have elongated limbs that help them cross gaps, reach fruit or other resources, test the firmness of support ahead, and in some cases, to brachiate. However, some species of lizard have reduced limb size that helps them avoid limb movement being obstructed by impinging branches.

Prehensile Tails

Many arboreal species, such as tree porcupines, chameleons, silky anteaters, spider monkeys, and possums, use prehensile tails to grasp branches. In the spider monkey and crested gecko, the tip of the tail has either a bare patch or adhesive pad, which provide increased friction.

Claws

Claws can be used to interact with rough substrates and re-orient the direction of forces the animal applies. This is what allows squirrels to climb tree trunks that are so large as to be essentially flat, from the perspective of such a small animal. However, claws can interfere with an animal's ability to grasp very small branches, as they may wrap too far around and prick the animal's own paw.

The silky anteater uses its prehensile tail as a third arm for stabilization and balance, while its claws help better grasp and climb onto branches.

Adhesion

Adhesion is an alternative to claws, which works best on smooth surfaces. Wet adhesion is common in tree frogs and arboreal salamanders, and functions either by suction or by capillary adhesion. Dry adhesion is best typified by the specialized toes of geckos, which use van der Waals forces to adhere to many substrates, even glass.

Gripping

Frictional gripping is used by primates, relying upon hairless fingertips. Squeezing the branch between the fingertips generates frictional force that holds the animal's hand to the branch. However, this type of grip depends upon the angle of the frictional force, thus upon the diameter of the branch, with larger branches resulting in reduced gripping ability. Animals other than primates that use gripping in climbing include the chameleon, which has mitten-like grasping feet, and many birds that grip branches in perching or moving about.

Reversible Feet

To control descent, especially down large diameter branches, some arboreal animals such as squirrels have evolved highly mobile ankle joints that permit rotating the foot into a 'reversed' posture. This allows the claws to hook into the rough surface of the bark, opposing the force of gravity.

Low Center of Mass

Many arboreal species lower their center of mass to reduce pitching and toppling movement when climbing. This may be accomplished by postural changes, altered body proportions, or smaller size.

Small Size

Small size provides many advantages to arboreal species: such as increasing the relative size of branches to the animal, lower center of mass, increased stability, lower mass (allowing movement on smaller branches), and the ability to move through more cluttered habitat. Size relating to weight affects gliding animals such as the reduced weight per snout-vent length for 'flying' frogs.

Hanging under Perches

Some species of primate, bat, and all species of sloth achieve passive stability by hanging beneath the branch. Both pitching and tipping become irrelevant, as the only method of failure would be losing their grip.

The gecko's toes adhere to surfaces via dry adhesion, to allow them to stay firmly attached to a branch or even a flat wall.

Behavioral Specializations

Arboreal species have behaviors specialized for moving in their habitats, most prominently in terms of posture and gait. Specifically, arboreal mammals take longer steps, extend their limbs further forwards and backwards during a step, adopt a more 'crouched' posture to lower their center of mass, and use a diagonal sequence gait.

Ecological Consequences

Arboreal locomotion allows animals access to different resources, depending upon their abilities. Larger species may be restricted to larger-diameter branches that can support their weight, while smaller species may avoid competition by moving in the narrower branches.

Climbing without Trees

Many animals climb in other habitats, such as in rock piles or mountains, and in those habitats, many of the same principles apply due to inclines, narrow ledges, and balance issues. However, less research has been conducted on the specific demands of locomotion in these habitats.

Perhaps the most exceptional of the animals that move on steep or even near vertical rock faces by careful balancing and leaping are the various types of mountain dwelling caprid such as the Barbary sheep, markhor, yak, ibex, tahr, rocky mountain goat, and chamois. Their adaptations may include a soft rubbery pad between their hooves for grip, hooves with sharp keratin rims for lodging in small footholds, and prominent dew claws. The snow leopard, being a predator of such mountain caprids, also has spectacular balance and leaping abilities; being able to leap up to ≈17m (~50 ft). Other balancers and leapers include the mountain zebra, mountain tapir, and hyraxes.

Brachiation

Arboreal snails use their sticky slime to help in climbing up trees, since they lack limbs to do so.

Brachiation is a specialized form of arboreal locomotion, used by primates to move very rapidly while hanging beneath branches. Arguably the epitome of arboreal locomotion, it involves swinging with the arms from one handhold to another. Only a few species are brachiators, and all of these are primates; it is a major means of locomotion among spider monkeys and gibbons, and is occasionally used by female orangutans. Gibbons are the experts of this mode of locomotion, swinging from branch to branch distances of up to 15 m (50 ft), and traveling at speeds of as much as 56 km/h (35 mph).

Gliding and Parachuting

To bridge gaps between trees, many animals such as the flying squirrel have adapted membranes, such as patagia for gliding flight. Some animals can slow their descent in the air using a method known as parachuting, such as *Rhacophorus* (a "flying frog" species) that has adapted toe membranes allowing it to fall more slowly after leaping from trees.

Limbless Climbing

Many species of snake are highly arboreal, and some have evolved specialized musculature for this habitat. While moving in arboreal habitats, snakes move slowly along bare branches using a specialized form of concertina locomotion, but when secondary branches emerge from the branch being moved on, snakes use lateral undulation, a much faster mode. As a result, snakes perform best on small perches in cluttered environments, while limbed organisms seem to do best on large perches in uncluttered environments.

Arboreal Animals

Many species of animals are arboreal, far too many to list individually. This list is of prominently or predominantly arboreal species and higher taxa.

- Primates,
- Cats,
- Brushtail possums,

- Opossums,
- Sloths,
- Anteaters,
- Treeshrews,
- Goats,
- Colugos,
- Kinkajous,
- Viverrids,
- Many rodents,
- Parrots,
- Geckos,
- Chameleons,
- Many other lizards,
- Mambas,
- Brown Tree Snakes,
- Many other snakes,
- Stick insects,
- Many other arthropods,
- Tree snails,
- Koalas.

Terrestrial Locomotion

An example of terrestrial locomotion. A horse - an erect-stanced unguligrade quadruped - with a galloping gait. An animation of photos by Eadweard Muybridge.

Terrestrial locomotion has evolved as animals adapted from aquatic to terrestrial environments. Locomotion on land raises different problems than that in water, with reduced friction being replaced by the effects of gravity.

There are three basic forms of locomotion found among terrestrial animals:

- Legged - Moving by using appendages.
- Limbless locomotion - moving without legs, primarily using the body itself as a propulsive structure.
- Rolling - rotating the body over the substrate.

Legged Locomotion

Movement on appendages is the most common form of terrestrial locomotion, it is the basic form of locomotion of two major groups with many terrestrial members, the vertebrates and the arthropods. Important aspects of legged locomotion are posture (the way the body is supported by the legs), the number of legs, and the functional structure of the leg and foot. There are also many gaits, ways of moving the legs to locomote, such as walking, running, or jumping.

Posture

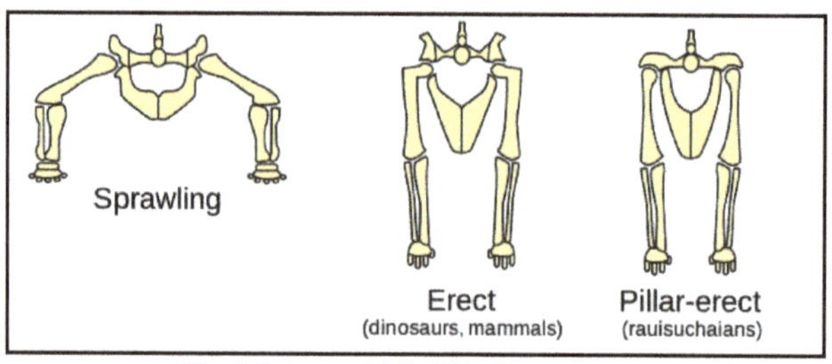

Hip joints and hindlimb postures.

Appendages can be used for movement in a number of ways. The posture, the way the body is supported by the legs, is an important aspect. There are three main ways in which vertebrates support themselves with their legs - sprawling, semi-erect, and fully erect. Some animals may use different postures in different circumstances, depending on the posture's mechanical advantages. Interestingly, there is no detectable difference in energetic cost between stances.

The "sprawling" posture is the most primitive, and is the original limb posture from which the others evolved. The upper limbs are typically held horizontally, while the lower limbs are vertical, though upper limb angle may be substantially increased in large animals. The body may drag along the ground, as in salamanders, or may be substantially elevated, as in monitor lizards. This posture is typically associated with trotting gaits, and the body flexes from side-to-side during movement to increase step length. All limbed reptiles and salamanders use this posture, as does the platypus and several species of frogs that walk. Unusual examples can be found among amphibious fish, such as the mudskipper, which drag themselves across land on their sturdy fins. Among the invertebrates, most arthropods—which includes the most diverse group of animals, the insects—have a stance best described as sprawling. There is also anecdotal evidence that some octopus species (such as the *Pinnoctopus* genus) can also drag themselves across land a short distance by hauling their body along by their tentacles (for example to pursue prey between rockpools) - there may be video evidence of this. The semi-erect posture is more accurately interpreted as an extremely elevated sprawling posture. This mode of locomotion is typically found in large lizards such as monitor lizards and tegus.

Mammals and birds typically have a fully erect posture, though each evolved it independently. In these groups the legs are placed beneath the body. This is often linked with the evolution of endothermy, as it avoids Carrier's constraint and thus allows prolonged periods of activity. The fully erect stance is not necessarily the "most-evolved" stance; evidence suggests that crocodilians evolved a semi-erect stance in their forelimbs from ancestors with fully erect stance as a result of adapting to a mostly aquatic lifestyle, though their hindlimbs are still held fully erect. For example, the mesozoic prehistoric crocodilian *Erpetosuchus* is believed to have had a fully erect stance and been terrestrial.

Number of Legs

The velvet worm (Onychophora)

The number of locomotory appendages varies much between animals, and sometimes the same animal may use different numbers of its legs in different circumstances. The best contender for unipedal movement is the springtail, which while normally hexapedal, hurls itself away from danger using its furcula, a tail-like forked rod that can be rapidly unfurled from the underside of its body.

A number of species move and stand on two legs, that is, they are bipedal. The group that is exclusively bipedal is the birds, which have either an alternating or a hopping gait. There are also a number of bipedal mammals. Most of these move by hopping – including the macropods such as kangaroos and various jumping rodents. Only a few mammals such as humans and the ground pangolin commonly show an alternating bipedal gait. Cockroaches and some lizards may also run on their two hind legs.

With the exception of the birds, terrestrial vertebrate groups with legs are mostly quadrupedal – the mammals, reptiles, and the amphibians usually move on four legs. There are many quadrupedal gaits. The most diverse group of animals on earth, the insects, are included in a larger taxon known as hexapods, most of which are hexapedal, walking and standing on six legs. Exceptions among the insects include praying mantises and water scorpions, which are quadrupeds with their front two legs modified for grasping, some butterflies such as the Lycaenidae (blues and hairstreaks) which use only four legs, and some kinds of insect larvae that may have no legs (e.g., maggots), or additional prolegs (e.g., caterpillars).

Spiders and many of their relatives move on eight legs – they are octopedal. However, some creatures move on many more legs. Terrestrial crustaceans may have a fair number - woodlice having fourteen legs. Also, as previously mentioned, some insect larvae such as caterpillars and sawfly larvae have up to five (caterpillars) or nine (sawflies) additional fleshy prolegs in addition to the six legs normal for insects. Some species of invertebrate have even more legs, the unusual velvet worm having stubby legs under the length of its body, with around several dozen pairs of legs. Centipedes have one pair of legs per body segment, with typically around 50 legs, but some species have over 200. The terrestrial animals with the most legs are the millipedes. They have two pairs of legs per body segment, with common species having between 80 and 400 legs overall – with the rare species *Illacme plenipes* having up to 750 legs. Animals with many legs typically move them in metachronal rhythm, which gives the appearance of waves of motion travelling forwards along their rows of legs.

Leg and Foot Structure

The legs of tetrapods, the main group of terrestrial vertebrates, have internal bones, with externally attached muscles for movement, and the basic form has three key joints: the shoulder joint, the knee joint, and the ankle joint, at which the foot is attached. Within this theme there is much variation in form. An alternative form of vertebrate 'leg' to the tetrapod leg is the fins found on amphibious fish. Also a few tetrapods, such as the macropods, have adapted their tails as additional locomotory appendages.

The basic form of the vertebrate foot has five toes, however some animals will have evolved fewer than this, and some early tetrapods had more; Acanthostega had eight toes. Feet have evolved many forms depending on the animal's needs. One key variation is where on the foot the animal's weight is placed. Most vertebrates—the amphibians, the reptiles, and some mammals such as humans and bears—are plantigrade, walking on the whole of the underside of the foot. Many mammals, such as cats and dogs are digitigrade, walking on their toes, the greater stride length allowing more speed. Digitigrade mammals are also often adept at quiet movement. Birds are also digitigrade. Some animals such as horses are unguligrade, walking on the tips of their toes. This even further increases their stride length and thus their speed. A few mammals are also known to walk on their knuckles, at least for their front legs. Knuckle-walking allows the foot (hand) to specialise for food gathering and/or climbing, as with the great apes and the extinct chalicotheres, or for swimming, as with the platypus. In animals where feet have evolved into functional hands, hand walking is also possible.

Among terrestrial invertebrates there are a number of leg forms. The arthropod legs are jointed and supported by hard external armor, with the muscles attached to the internal surface of this exoskeleton. The other group of legged terrestrial invertebrates, the velvet worms, have soft stumpy legs supported by a hydrostatic skeleton. The prolegs that some caterpillars have in addition to their six more-standard arthropod legs have a similar form to those of velvet worms, and suggest a distant shared ancestry.

Gaits

A jumping kangaroo.

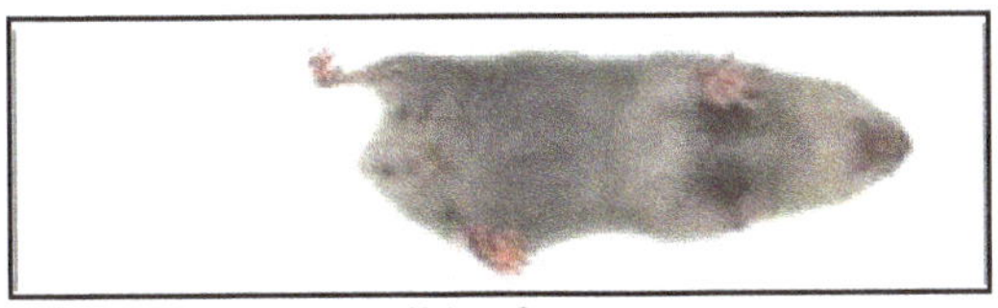

A walking hamster.

Animals show a vast range of gaits, the order that they place and lift their appendages in locomotion. Gaits can be grouped into categories according to their patterns of support sequence. For quadrupeds, there are three main categories: walking gaits, running gaits, and leaping gaits. In one system (relating to horses), there are 60 discrete patterns: 37 walking gaits, 14 running gaits, and 9 leaping gaits.

Walking is the most common gait, where some feet are on the ground at any given time, and found in almost all legged animals. In an informal sense, running is considered to occur when at some points in the stride all feet are off the ground in a moment of suspension. Technically, however, moments of suspension occur in both running gaits (such as trot) and leaping gaits (such as canter and gallop). Gaits involving one or more moments of suspension can be found in many animals, and compared to walking they are faster but more energetically costly forms of locomotion.

Animals will use different gaits for different speeds, terrain, and situations. For example, horses show four natural gaits, the slowest horse gait is the walk, then there are three faster gaits which, from slowest to fastest, are the trot, the canter, and the gallop. Animals may also have unusual gaits that are used occasionally, such as for moving sideways or backwards. For example, the main human gaits are bipedal walking and running, but they employ many other gaits occasionally, including a four-legged crawl in tight spaces.

In walking, and for many animals running, the motion of legs on either side of the body alternates, i.e. is out of phase. Other animals, such as a horse when galloping, or an inchworm, alternate between their front and back legs. An alternative to a gait that alternates between legs is hopping or saltation, where all legs move together. As a main means of locomotion, this is usually found in bipeds or semi-bipeds. Among the mammals saltation is commonly used among macropods (kangaroos and their relatives), jerboas, springhares, kangaroo rats, hopping mice, gerbils, and sportive lemurs. Certain tendons in kangaroo hind legs are very elastic, allowing kangaroos to effectively bounce along conserving energy from hop to hop, making hopping a very energy efficient way to move around in their nutrient poor environment. Saltation is also used by many small birds. Frogs and fleas also hop.

Most animals move in the direction of their head. However, there are some exceptions. Crabs move sideways, and naked mole rats, which live in tight tunnels underground, can move backward or forward with equal facility. Crayfish can move backward much faster than they can move forward.

Gait analysis is the study of gait in humans and other animals. This may involve videoing subjects with markers on particular anatomical landmarks and measuring the forces of their footfall using floor transducers (strain gauges). Skin electrodes may also be used to measure muscle activity.

Limbless Locomotion

A snail moves by slithering.

There are a number of terrestrial and amphibious limbless vertebrates and invertebrates. These animals, due to lack of appendages, use their bodies to generate propulsive force. These movements are sometimes referred to as "slithering" or "crawling", although neither are formally used in the scientific literature and the latter term is also used for some animals moving on all four limbs. All limbless animals come from cold-blooded groups; there are no endothermic limbless animals, i.e. there are no limbless birds or mammals.

Lower Body Surface

Where the foot is important to the legged mammal, for limbless animals the underside of the body is important. Some animals such as snakes or legless lizards move on their smooth dry underside. Other animals have various features that aid movement. Molluscs such as slugs and snails move on a layer of mucus that is secreted from their underside, reducing friction and protecting from injury when moving over sharp objects. Earthworms have small bristles (setae) that hook into the substrate and help them move. Some animals such as leeches have suction cups on either end of the body allowing two anchor movement.

Type of Movement

Some limbless animals, such as leeches, have suction cups on either end of their body, which allow them to move by anchoring the rear end and then moving forward the front end, which is then anchored and then the back end is pulled in, and so on. This is known as two-anchor movement. A legged animal, the inchworm, also moves like this, clasping with appendages at either end of its body.

Limbless animals can also move using pedal locomotary waves, rippling the underside of the body. This is the main method used by molluscs such as slugs and snails, and also large flatworms, some other worms, and even earless seals. The waves may move in the opposite direction to motion, known as retrograde waves, or in the same direction as motion, known as direct waves. Earthworms move by retrograde waves alternatively swelling and contracting down the length of their body, the swollen sections being held in place using setae. Aquatic molluscs such as limpets, which are sometimes out of the water, tend to move using retrograde waves. However terrestrial molluscs such as slugs and snails tend to use direct waves. Lugworms and seals also use direct waves.

Most snakes move using lateral undulation where a lateral wave travels down the snake's body in the opposite direction to the snake's motion and pushes the snake off irregularities in the ground. This mode of locomotion requires these irregularities to function. Another form of locomotion, rectilinear locomotion, is used at times by some snakes, especially large ones such as pythons and boa. Here large scales on the underside of the body, known as scutes are used to push backwards and downwards. This is effective on a flat surface and is used for slow, silent movement, such as when stalking prey. Snakes use concertina locomotion for moving slowly in tunnels, here the snake alternates in bracing parts of its body on it surrounds. Finally the caenophidian snakes use the fast and unusual method of movement known as sidewinding on sand or loose soil. The snake cycles through throwing the front part of its body in the direction of motion and bringing the back part of its body into line crosswise.

Rolling

The pangolin Manis temminckii in defensive position.

Although animals have never evolved wheels for locomotion, a small number of animals will move at times by rolling their whole body. Rolling animals can be divided into those that roll under the force of gravity or wind and those that roll using their own power.

Gravity or Wind Assisted

The web-toed salamander, a 10-centimetre (3.9 in) salamander, lives on steep hills in the Sierra Nevada mountains. When disturbed or startled it coils itself up into a ball, often causing it to roll downhill.

The pebble toad (*Oreophrynella nigra*) lives atop tepui in the Guiana highlands of South America. When threatened, often by tarantulas, it rolls into ball, and typically being on an incline, rolls away under gravity like a loose pebble.

Namib wheeling spiders (*Carparachne spp.*), found in the Namib desert, will actively roll down sand dunes. This action can be used to successfully escape predators such as the *Pompilidae* tarantula wasps, which lay their eggs in a paralyzed spider for their larvae to feed on when they hatch.

The spiders flip their body sideways and then cartwheel over their bent legs. The rotation is fast, the golden wheel spider (*Carparachne aureoflava*) moving up to 20 revolutions per second, moving the spider at 1 metre per second.

Coastal tiger beetle larvae when threatened can flick themselves into the air and curl their bodies to form a wheels, which the wind blows, often uphill, as far as 25m and as fast as 11kph. The also may have some ability to steer themselves in this state.

Pangolins, a type of mammal covered in thick scales, roll into a tight ball when threatened. Pangolins have been reported to roll away from danger, by both gravity and self-powered methods. A pangolin in hill country in Sumatra, to flee from the researcher, ran to the edge of a slope and curled into a ball to roll down the slope, crashing through the vegetation, and covering an estimated 30 metres or more in 10 seconds.

Self-powered

Caterpillars of the Mother-Of-Pearl Moth, *Pleuroptya ruralis*, when attacked, will touch their heads to their tails and roll backwards, up to 5 revolutions at about 40 cm per second, which is about 40 times its normal speed.

Nannosquilla decemspinosa, a species of long-bodied, short-legged mantis shrimp, lives in shallow sandy areas along the Pacific coast of Central and South America. When stranded by a low tide the 3 cm stomatopod lies on its back and performs backwards somersaults over and over. The animal moves up to 2 meters at a time by rolling 20–40 times, with speeds of around 72 revolutions per minute. That is 1.5 body lengths per second (3.5 cm/s). Researchers estimate that the stomatopod acts as a true wheel around 40% of the time during this series of rolls. The remaining 60% of the time it has to "jumpstart" a roll by using its body to thrust itself upwards and forwards.

Pangolins have also been reported to roll away from danger by self-powered methods. Witnessed by a lion researcher in the Serengeti in Africa, a group of lions surrounded a pangolin, but could not get purchase on it when it rolled into a ball, and so the lions sat around it waiting and dozing. Surrounded by lions, it would unroll itself slightly and give itself a push to roll some distance, until by doing this multiple times it could get far enough away from the lions to be safe. Moving like this would allow a pangolin to cover distance while still remaining in a protective armoured ball.

Limits and Extremes

The fastest terrestrial animal is the cheetah, which can attain maximal sprint speeds of approximately 104 km/h (64 mph). The fastest running lizard is the Black Iguana, which has been recorded moving at speed of up to 34.9 km/h (21.7 mph).

Animal Locomotion on the Water Surface

Animal locomotion on the surface layer of water is the study of animal locomotion in the case of small animals that live on the surface layer of water, relying on surface tension to stay afloat.

Water striders can move on the surface of water.

There are two types of animal locomotion on water, determined by the ratio of the animal's weight to the water's surface tension: those whose weight is supported by the surface tension at rest, and can therefore easily remain on the water's surface without much exertion, and those whose weight is not supported by the water's surface tension at rest, and must therefore exert additional motion in a direction parallel to the water's surface in order to remain above it. A creature such as the basilisk lizard, often dubbed the 'Jesus lizard', has a weight which is larger than the surface tension can support, and is widely known for running across the surface of water. Another example, the western grebe, performs a mating ritual that includes running across the surface of water.

Surface living animals such as the water strider typically have hydrophobic feet covered in small hairs that prevent the feet from breaking the surface and becoming wet. Another insect known to walk on the water surface is the ant species *Polyrhachis sokolova*. The pygmy gecko (*Coleodactylus amazonicus*), due to its small size and hydrophobic skin is also able to walk on the water surface.

According to biophysicist David L. Hu, there are at least 342 species of water striders. As striders increase in size, their legs become proportionately longer, with *Gigantometra gigas* having a length of over 20 cm requiring a surface tension force of about 40 millinewtons.

Water striders generate thrust by shedding vortices in the water: a series of "U"-shaped vortex filaments is created during the power stroke. The two free ends of the "U" are attached to the water surface. These vortices transfer enough (backward) momentum to the water to propel the animal forwards (note that some momentum is transferred by capillary waves.

Meniscus Climbing

To pass from the water surface to land, a water-walking insect must contend with the slope of the meniscus at the water's edge. Many such insects are unable to climb this meniscus using their usual propulsion mechanism.

David Hu and coworker John W. M. Bush have shown that such insects climb meniscuses by assuming a fixed body posture. This deforms the water surface and generates capillary forces that propels the insect up the slope without moving its appendages.

Hu and Bush conclude that meniscus climbing is an unusual means of propulsion in that the insect propels itself in a quasi-static configuration, without moving its appendages. Biolocomotion is generally characterized by the transfer of muscular strain energy to the kinetic and gravitational potential energy of the creature, and the kinetic energy of the suspending fluid. In contrast, meniscus climbing has a different energy pathway: by deforming the free surface, the insect converts muscular strain to the surface energy that powers its ascent.

Marangoni Propulsion

Many insects, including some terrestrial insects, can release a surfactant and propel themselves using the Marangoni effect. Hu and Bush report that *Microvelia* can attain a peak speed of 17 cm/s, which is twice its peak walking speed, using Marangoni propulsion.

Marangoni propulsion by a wetting arthropod is precisely analogous to a soap boat but the situation for insects such as water striders is more complex. Hu and Bush state that "for nonwetting arthropods, the transfer of chemical to kinetic energy is more subtle, as the Marangoni stress must be communicated across the creature's complex surface layer".

Sailing

Elella, the by-the-wind sailor, is a cnidarian with no means of propulsion other than sailing. A small rigid sail projects into the air and catches the wind. *Velella* sails always align along the direction of the wind where the sail may act as an aerofoil, so that the animals tend to sail downwind at a small angle to the wind.

Undulatory Locomotion

Snakes primarily rely on undulatory locomotion to move through a wide range of environments.

Undulatory locomotion is the type of motion characterized by wave-like movement patterns that act to propel an animal forward. Examples of this type of gait include crawling in snakes, or swimming in the lamprey. Although this is typically the type of gait utilized by limbless animals, some creatures with limbs, such as the salamander, choose to forgo use of their legs in certain environments and exhibit undulatory locomotion. This movement strategy is important to study in order to create novel robotic devices capable of traversing a variety of environments.

Environmental Interactions

In limbless locomotion, forward locomotion is generated by propagating flexural waves along the length of the animal's body. Forces generated between the animal and surrounding environment lead to a generation of alternating sideways forces that act to move the animal forward. These forces generate thrust and drag.

Hydrodynamics

Simulation predicts that thrust and drag are dominated by viscous forces at low Reynolds numbers and inertial forces at higher Reynolds numbers. When the animal swims in a fluid, two main forces are thought to play a role:

- Skin Friction: Generated due to the resistance of a fluid to shearing and is proportional to speed of the flow. This dominates undulatory swimming in spermatozoa and the nematode.
- Form Force: Generated by the differences in pressure on the surface of the body and it varies with the square of flow speed.

At low Reynolds number ($Re \sim 10^0$), skin friction accounts for nearly all of the thrust and drag. For those animals which undulate at intermediate Reynolds number ($Re \sim 10^1$), such as the Ascidian larvae, both skin friction and form force account for the production of drag and thrust. At high Reynolds number ($Re \sim 10^2$), both skin friction and form force act to generate drag, but only form force produces thrust.

Kinematics

In animals that move without use of limbs, the most common feature of the locomotion is a rostral to caudal wave that travels down their body. However, this pattern can change based on the particular undulating animal, the environment, and the metric in which the animal is optimizing (i.e. speed, energy, etc.). The most common mode of motion is simple undulations in which lateral bending is propagated from head to tail.

Snakes can exhibit 5 different modes of terrestrial locomotion: (1) lateral undulation, (2) side-winding, (3) concertina, (4) rectilinear, and (5) slide-pushing. Lateral undulation closely resembles the simple undulatory motion observed in many other animals such as in lizards, eels and fish, in which waves of lateral bending propagate down the snakes body.

The American eel typically moves in an aquatic environment, though it can also move on land for short periods of time. It is able to successfully move about in both environments by producing traveling waves of lateral undulations. However, differences between terrestrial and aquatic locomotor strategy suggest that the axial musculature is being activated differently, In terrestrial locomotion, all points along the body move on approximately the same path and, therefore, the lateral displacements along the length of the eel's body is approximately the same. However, in aquatic locomotion, different points along the body follow different paths with increasing lateral amplitude more posteriorly. In general, the amplitude of the lateral undulation and angle of intervertebral flexion is much greater during terrestrial locomotion than that of aquatic.

Musculoskeletal System

Perch filets showing myomere structure.

Muscle Architecture

A typical characteristic of many animals that utilize undulatory locomotion is that they have segmented muscles, or blocks of myomeres, running from their head to tails which are separated by connective tissue called myosepta. In addition, some segmented muscle groups, such as the lateral hypaxial musculature in the salamander are oriented at an angle to the longitudinal direction. For these obliquely oriented fibers the strain in the longitudinal direction is greater than the strain in the muscle fiber direction leading to an architectural gear ratio greater than 1. A higher initial angle of orientation and more dorsoventral bulging produces a faster muscle contraction but results in a lower amount of force production. It is hypothesized that animals employ a variable gearing mechanism that allows self-regulation of force and velocity to meet the mechanical demands of the contraction. When a pennate muscle is subjected to a low force, resistance to width changes in the muscle cause it to rotate which consequently produce a higher architectural gear ratio (AGR) (high velocity). However, when subject to a high force, the perpendicular fiber force component overcomes the resistance to width changes and the muscle compresses producing a lower AGR (capable of maintaining a higher force output).

Most fishes bend as a simple, homogenous beam during swimming via contractions of longitudinal red muscle fibers and obliquely oriented white muscle fibers within the segmented axial musculature. The fiber strain (εf) experienced by the longitudinal red muscle fibers is equivalent to the longitudinal strain (εx). The deeper white muscle fibers fishes show diversity in arrangement. These fibers are organized into cone-shaped structures and attach to connective tissue sheets known as myosepta; each fiber shows a characteristic dorsoventral (α) and mediolateral (φ) trajectory. The segmented architecture theory predicts that, $\varepsilon x > \varepsilon f$. This phenomenon results in an architectural gear ratio, determined as longitudinal strain divided by fiber strain ($\varepsilon x / \varepsilon f$), greater than one and longitudinal velocity amplification; furthermore, this emergent velocity amplification may be augmented by variable architectural gearing via mesolateral and dorsoventral shape changes, a pattern seen in pennate muscle contractions. A red-to-white gearing ratio (red εf / white εf) captures the combined effect of the longitudinal red muscle fiber and oblique white muscle fiber strains.

Simple bending behavior in homogenous beams suggests ε increases with distance from the neutral axis (z). This poses a problem to animals, such as fishes and salamanders, which undergo undulatory movement. Muscle fibers are constrained by the length-tension and force-velocity curves. Furthermore, it has been hypothesized that muscle fibers recruited for a particular task must operate within an optimal range of strains (ε) and contractile velocities to generate peak force and power respectively. Non-uniform ε generation during undulatory movement would force differing muscle fibers recruited for the same task to operate on differing portions of the length-tension and force-velocity curves; performance would not be optimal. Alexander predicted that the dorsoventral (α) and mediolateral (φ) orientation of the white fibers of the fish axial musculature may allow more uniform strain across varying mesolateral fiber distances. Unfortunately, the white muscle fiber musculature of fishes is too complex to study uniform strain generation; however, Azizi *et al.* studied this phenomenon using a simplified salamander model.

Siren lacertian, an aquatic salamander, utilizes swimming motions similar to the aforementioned fishes yet contains hypaxial muscle fibers (which generate bending) characterized by a simpler organization. The hypaxial muscle fibers of S. lacertian are obliquely oriented, but have a near zero mediolateral (φ) trajectory and a constant doroslateral (α) trajectory within each segment. Therefore, the effect of doroslateral (α) trajectory and the distance between a given hypaxial muscle layer and the neutral axis of bending (z) on muscle fiber strain (ε) can be studied.

Azizi *et al.* found that longitudinal contractions of the constant volume hypaxial muscles were compensated by an increase in the dorsoventral dimensions. Bulging was accompanied by fiber rotation as well as an increase in both α hypaxial fiber trajectory and architectural gear ratio (AGR), a phenomenon also seen in pennate muscle contractions. Azizi *et al.* constructed a mathematical model to predict the final hypaxial fiber angle, AGR and dorsoventral height, where: λx = longitudinal extension ratio of the segment (portortion of final longitudinal length after contraction to initial longitudinal length), β = final fiber angle, γ = initial fiber angle, f = initial fiber length, and εx and εf = longitudinal and fiber strain respectively.

- $\lambda x = \varepsilon x + 1$
- $\lambda f = \varepsilon f + 1$
- $\varepsilon x = \Delta L / Linital$
- Architectural gear ratio = $\varepsilon x / \varepsilon f = [\lambda f(\cos\beta / \cos\gamma) - 1)] / (\lambda f - 1)$
- $\beta = \sin-1\,(y2 / \lambda ff)$

This relationship shows that AGR increase with an increase in fiber angle from γ to β. In addition, final fiber angle (β) increases with dorsolateral bulging (y) and fiber contraction, but decreases as a function of initial fiber length.

The application of the latter conclusions can be seen in S. lacertian. This organism undulates as a homogenous beam (just as in fishes) during swimming; thus the distance of a muscle fiber from the neutral axis (z) during bending must be greater for external oblique muscle layers (EO) than internal oblique muscle layers (IO). The relationship between the strains (ε) experienced by the EO and IO and their respective z values is given by the following equation: where εEO and εIO =

strain of the external and internal oblique muscle layers, and zEO and zIO = distance of the external and internal oblique muscle layers respectively from the neutral axis.

$$\varepsilon EO = \varepsilon IO\ (zEO / zIO)$$

Via this equation, we see that z is directly proportional to ε; the strain experienced by the EO exceeds that of the IO. Azizi *et al.* discovered that the initial hypaxial fiber α trajectory in the EO is greater than that of the IO. Because initial α trajectory is proportional to the AGR, the EO contracts with a greater AGR than the IO. The resulting velocity amplification allows both layers of muscles to operate at similar strains and shortening velocities; this enables the EO and IO to function on comparable portions of the length-tension and force-velocity curves. Muscles recruited for a similar task ought to operate at similar strains and velocities to maximize force and power output. Therefore, variability in AGR within the hypaxial musculature of the Siren lacertian counteracts varying mesolateral fiber distances and optimizes performance. Azizi *et al.* termed this phenomenon as fiber strain homogeneity in segmented musculature.

Muscle Activity

In addition to a rostral to caudal kinematic wave that travels down the animals body during undulatory locomotion, there is also a corresponding wave of muscle activation that travels in the rostro-caudal direction. However, while this pattern is characteristic of undulatory locomotion, it too can vary with environment.

American Eel

Aquatic Locomotion: Electromyogram (EMG) recordings of the American eel reveal a similar pattern of muscle activation during aquatic movement as that of fish. At slow speeds only the most posterior end of the eel's muscles are activated with more anterior muscle recruited at higher speeds. As in many other animals, the muscles activate late in the lengthening phase of the muscle strain cycle, just prior to muscle shortening which is a pattern believed to maximize work output from the muscle.

Terrestrial Locomotion: EMG recordings show a longer absolute duration and duty cycle of muscle activity during locomotion on land. Also, the absolute intensity is much higher while on land which is expected from the increase in gravitational forces acting on the animal. However, the intensity level decreases more posteriorly along the length of the eel's body. Also, the timing of muscle activation shifts to later in the strain cycle of muscle shortening.

Energetics

Animals with elongated bodies and reduced or no legs have evolved differently from their limbed relatives. In the past, some have speculated that this evolution was due to a lower energetic cost associated with limbless locomotion. The biomechanical arguments used to support this rationale include that (1) there is no cost associatied with the vertical displacement of the center of mass typically found with limbed animals, (2) there is no cost associated with accelerating or decelerating limbs, and (3) there is a lower cost for supporting the body. This hypothesis has been studied further by examining the oxygen consumption rates in the snake during different modes of locomotion: lateral undulation, concertina, and sidewinding. The *net cost of transport* (NCT), which indicates the amount of energy required to move a unit of mass a given distance, for a snake moving with a lateral undulatory gait is identical to that of a limbed lizard with the same mass. However, a snake utilizing concertina locomotion produces a much higher net cost of transport, while sidewinding actually produces a lower net cost of transport. Therefore, the different modes of locomotion are of primary importance when determining energetic cost. The reason that lateral undulation has the same energetic efficiency as limbed animals and not less, as hypothesized earlier, might be due to the additional biomechanical cost associated with this type of movement due to the force needed to bend the body laterally, push its sides against a vertical surface, and overcome sliding friction.

Neuromuscular System

Intersegmental Coordination

Wavelike motor pattern typically arise from a series of coupled segmental oscillator. Each segmental oscillator is capable of producing a rhythmic motor output in the absence of sensory feedback. One such example is the half center oscillator which consist of two neurons that are mutually inhibitory and produce activity 180 degrees out of phase. The phase relationships between these oscillators are established by the emergent properties of the oscillators and the coupling between them. Forward swimming can by accomplished by a series of coupled oscillators in which the anterior oscillators have a shorter endogenous frequency than the posterior oscillators. In this case, all oscillators will be driven at the same period but the anterior oscillators will lead in phase. In addition, the phase relations can be established by asymmetries in the couplings between oscillators or by sensory feedback mechanisms.

- Leech

The leech moves by producing dorsoventral undulations. The phase lags between body segments is about 20 degrees and independent of cycle period. Thus, both hemisegments of the oscillator fire synchronously to produce a contraction. Only the ganglia rostral to the midpoint are capable of producing oscillation individually. There is U-shaped gradient in endogenous segment oscillation as well with the highest oscillations frequencies occurring near the middle of the animal. Although the couplings between neurons spans six segments in both the anterior and posterior direction, there are asymmetries between the various interconnections because the oscillators are active at three different phases. Those that are active in the 0 degree phase project only in the descending direction while those projecting in the ascending direction are active at 120 degrees or 240 degrees. In addition, sensory feedback from the environment may contribute to resultant phase lag.

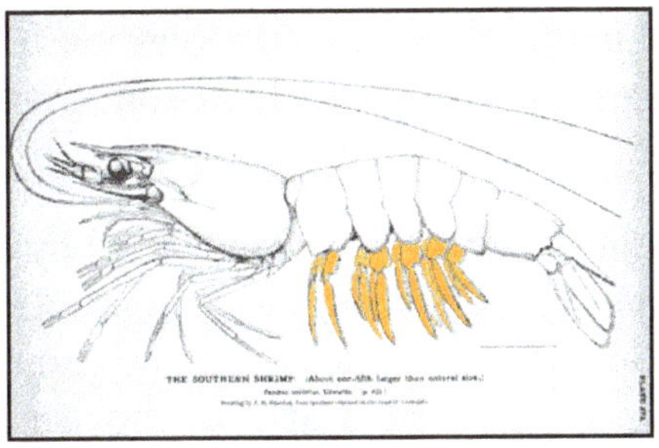

Pleopods

- Lamprey

The lamprey moves using lateral undulation and consequently left and right motor hemisegments are active 180 degrees out of phase. Also, it has been found that the endogenous frequency of the more anterior oscillators is higher than that of the more posterior ganglia. In addition, inhibitory interneurons in the lamprey project 14-20 segments caudally but have short rostral projections. Sensory feedback may be important for appropriately responding to perturbations, but seems to be less important for maintaince of appropriate phase relations.

Robotics

Based on biologically hypothesized connections of the central pattern generator in the salamander, a robotic system has been created which exhibits the same characteristics of the actual animal. Electrophysiology studies have shown that stimulation of the mesencephalic locomotor region (MLR) located in the brain of the salamander produce different gaits, swimming or walking, depending on intensity level. Similarly, the CPG model in the robot can exhibit walking at low levels of tonic drive and swimming at high levels of tonic drive. The model is based on the four assumptions that:

- Tonic stimulation of the body CPG produces spontaneous traveling waves. When the limb CPG is activated it overrides the body CPG.
- The strength of the coupling from the limb to the body CPG is stronger than that from body to limb.
- Limb oscillators saturate and stop oscillating at higher tonic drives.
- Limb oscillators have lower intrinsic frequencies than body CPGs at the same tonic drive.

This model encompasses the basic features of salamander locomotion.

Comparative Foot Morphology

Comparative foot morphology involves comparing the form of distal limb structures of a variety of terrestrial vertebrates. Understanding the role that the foot plays for each type of organism must take account of the differences in body type, foot shape, arrangement of structures, loading conditions and other variables. However, similarities also exist among the feet of many different terrestrial vertebrates. The paw of the dog, the hoof of the horse, the *manus* (foot) and *pes* (foot) of the elephant, and the foot of the human all share some common features of structure, organization and function. Their foot structures function as the load-transmission platform which is essential to balance, standing and types of locomotion (such as walking, trotting, galloping and running).

The discipline of biomimetics applies the information gained by comparing the foot morphology of a variety of terrestrial vertebrates to human-engineering problems. For instance, it may provide insights that make it possible to alter the foot's load transmission in people who wear an external orthosis because of paralysis from spinal-cord injury, or who use a prosthesis following the diabetes-related amputation of a leg. Such knowledge can be incorporated in technology that improves a person's balance when standing; enables them to walk more efficiently, and to exercise; or otherwise enhances their quality of life by improving their mobility.

Structure

Limb and foot structure of representative terrestrial vertebrates:

Variability in Scaling and Limb Coordination

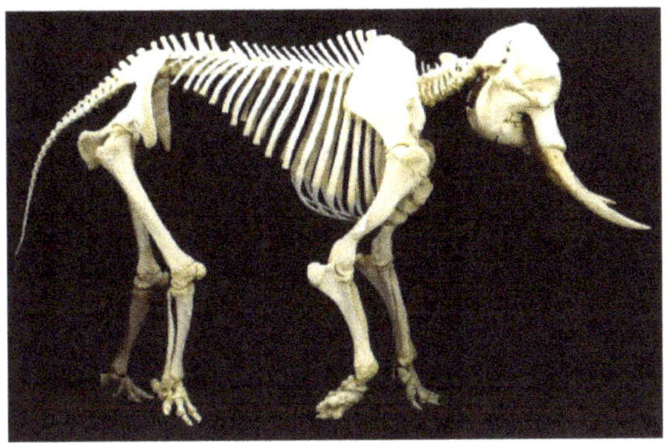

Elephant skeleton

There is considerable variation in the scale and proportions of body and limb, as well as the nature of loading, during standing and locomotion both among and between quadrupeds and bipeds. The anterior-posterior body mass distribution varies considerably among mammalian quadrupeds, which affects limb loading. When standing, many terrestrial quadrupeds support more of their weight on their forelimbs rather than their hind limbs; however, the distribution of body mass and limb loading changes when they move. Humans have a lower-limb mass that is greater than their upper-limb mass. The hind limbs of the dog and horse have a slightly greater mass than the forelimbs, whereas the elephant has proportionally longer limbs. The elephant's forelimbs are longer than its hind limbs.

In the horse and dog, the hind limbs play an important role in primary propulsion. The legged locomotion of humans generally distributes an equal loading on each lower limb. The locomotion of the elephant (which is the largest terrestrial vertebrate) displays a similar loading distribution on its hind limbs and forelimbs. The walking and running gaits of quadrupeds and bipeds show differences in the relative phase of the movements of their forelimbs and hind limbs, as well as of their right-side limbs versus their left-side limbs. Many of the aforementioned variables are connected with differences in the scaling of body and limb dimension as well as in patterns of limb coordination and movement. However, little is understood concerning the functional contribution of the foot and its structures during the weight-bearing phase. The comparative morphology of the distal limb and foot structure of some representative terrestrial vertebrates reveals some interesting similarities.

Columnar Organization of Limb Structures

Even many terrestrial vertebrates exhibit differences in the scaling of limb dimension, limb coordination and magnitude of forelimb-hind limb loading, in the dog, horse and elephant the structure of the distal forelimb is similar to that of the distal hind limb. In the human, the structures of the hand are generally similar in shape and arrangement to those of the foot. Terrestrial vertebrate quadrupeds and bipeds generally possess distal limb and foot endoskeleton structures that are aligned in series, stacked in a relatively vertical orientation and arranged in a quasi-columnar fashion in the extended limb. In the dog and horse, the bones of the proximal limbs are oriented vertically, whereas the distal limb structures of the ankle and foot have an angulated orientation. In humans and elephants, a vertical-column orientation of the bones in the limbs and feet is also evident for associated skeletal muscle-tendon units. The horse's foot contains an external nail (hoof) oriented about the perimeter in the shape of a semicircle. The underlying bones are arranged in a semi-vertical orientation. The dog's paw similarly contains bones arranged in a semi-vertical orientation.

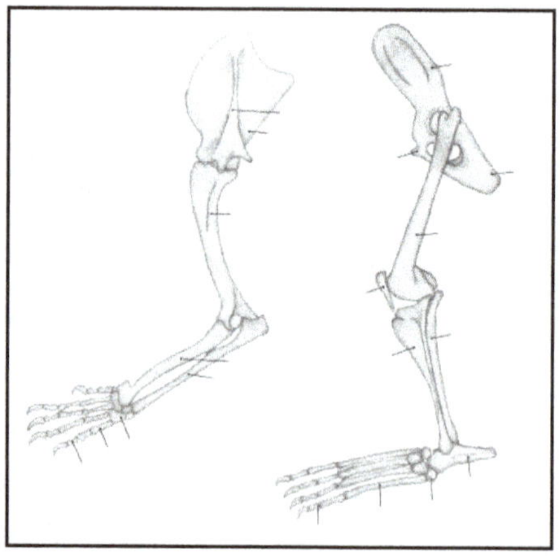

Limb skeleton of a lion, an example of an angulated bony column.

In the human and the elephant, the column orientation of the foot complex is replaced in humans by a plantigrade orientation, and in elephants by a semi-plantigrade alignment of the hind limb foot structure. This difference in orientation in the foot bones and joints of humans and elephants helps them to adapt to variations in the terrain.

Distal Cushion

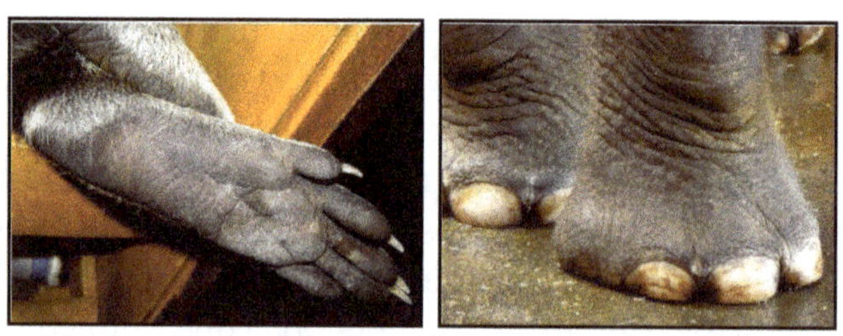

Distal cushions on the foot of a raccoon and an elephant.

Many representative terrestrial vertebrates possess a distal cushion on the under-surface of the foot. The dog's paw contains a number of visco-elastic pads oriented along the middle and distal foot. The horse possesses a centralized digital pad known as the frog, which is located at the distal aspect of the foot and surrounded by the hoof. Humans possess a tough fibro and elastic pad of fat that is anchored to the skin and bone of the rear portion of the foot.

The foot of the elephant possesses what is perhaps one of the most unusual distal cushions found in vertebrates. The forefoot (*manus*) and hindfoot (*pes*) contain huge pads of fat that are scaled to cope with the massive loadings imposed by the largest terrestrial vertebrate. In addition, a cartilage-like projection (prepollex in the forelimb and prehallux in the hind limb) appears to anchor the distal cushion to the bones of the elephant's foot.

The distal cushions of all these organisms (dog, horse, human and elephant) are dynamic structures during locomotion, alternating between phases of compression and expansion; it has been suggested that these structures thereby reduce the loads experienced by the skeletal system.

Organization

Arrangement of foot structures:

Because of the wide variety in body types, scaling and morphology of the distal limbs of terrestrial vertebrates, there exists a degree of controversy concerning the nature and organization of foot structures. One organizational approach to understanding foot structures makes distinctions regarding their regional anatomy. The foot structures are divided into segments from proximal to distal and are grouped according to similarity in shape, dimension and function. In this approach, the foot may be described in three segments: as the hindfoot, midfoot and forefoot.

The hindfoot is the most proximal and posterior portion of the foot. Functionally, the structures contained in this region are typically robust, possessing a larger size and girth than the other structures of the foot. The structures of the hindfoot are usually adapted for transmitting large loads between the proximal and distal aspects of the limb when the foot contacts the ground. This is apparent in the human and elephant foot, where the hindfoot undergoes greater loading during initial contact in many forms of locomotion. The hindfoot structures of the dog and horse are located relatively proximally compared to the elephant and human foot.

The midfoot is the intermediate portion of the foot between the hindfoot and forefoot. The structures in this region are intermediate in size, and typically transmit loads from the hindfoot to the forefoot. The human transverse tarsal joint of the midfoot transmits forces from the subtalar joint in the hindfoot to the forefoot joints (metatarsophalangeal and interphalangeal) and associated bones (metatarsals and phalanges). The midfoot of the dog, horse and elephant contains similar intermediate structures having similar functions to those of the human midfoot.

The forefoot represents the most distal portion of the foot. In the human and elephant, the bone structures contained in this region are generally longer and narrower. The structures of the forefoot play a role in providing leverage for terminal stance propulsion and load transfer.

Function

Load transmission of the foot in representative terrestrial vertebrates:

Dog Paw

The paw of the dog has a digitigrade orientation. The vertical columnar orientation of the proximal bones of the limbs, which articulate with distal foot structures that are arranged in quasi-vertical columnar orientation, is well-aligned to transmit loadings during weight-bearing contact of the skeleton with the ground. The angled orientation of the elongated metatarsal and the digits extends the area available for storing and releasing mechanical energy in the muscle tendon units originating proximally to the ankle joint and terminating at the distal aspect of the foot bones. When muscle tendon units lengthen, the load strain facilitates mechanical activity. These muscle tendon unit structures appear well designed to aid in the ground-reaction transmission of forces that is essential for locomotion. In addition, the pads of the distal paw appear to allow load attenuation, by enhancing shock absorption during the paw's contact with the ground.

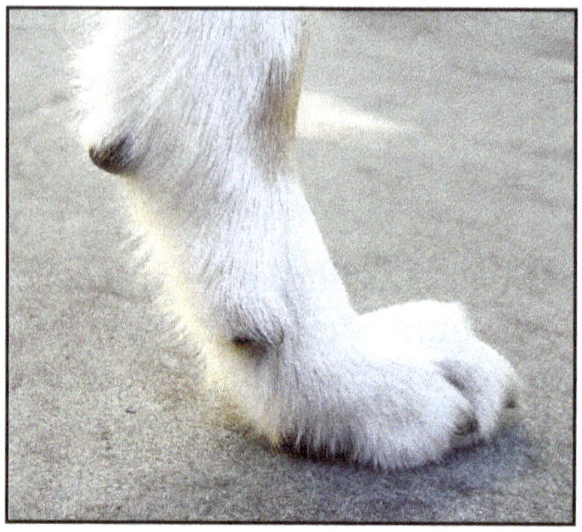

Dog paw

Horse Foot

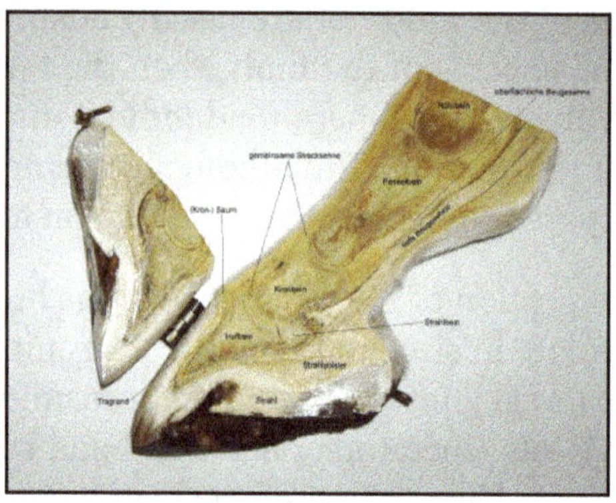

Section of a horse foot

The horse's foot is in an unguligrade orientation. The columnar orientation of bones and connective tissue is similarly well-aligned to transmit loads during the weight-bearing phase of locomotion. The thick keratinized and semicircular hoof changes shape during loading and unloading. Similarly, the cushioned frog situated centrally at the rear ends of the hoof undergoes compression during loading, and expansion when unloaded. Together, the hoof and cushioned frog structures may work in concert with hoof capsule to provide shock absorption. The horse hoof also acts dynamically during loading, which may cushion the endoskeleton from high loads that would otherwise produce critical deformation.

Elephant Foot

The hind limb and foot of the elephant are oriented semi-plantigrade, and closely resemble the structure and function of the human foot. The tarsals and metapodials are arranged so as to form an arch, similarly to the human foot. The six toes of each foot of the elephant are enclosed in a flexible sheath of skin. Similar to the dog's paw, the elephant's phalanges are oriented in a downward direction. The distal phalanges of the elephant do not directly touch the ground, and are attached to the respective nail/hoof. Distal cushions occupy the spaces between the muscle tendon units and ligaments within the hindfoot, midfoot and forefoot bones on the plantar surface. The distal cushion is highly innervated by sensory structures (Meissner's and Pacinian corpuscles), making the distal foot one of the most sensitive structures of the elephant (more so than its trunk). The cushions of the elephant's foot respond to the requirement to store and absorb mechanical loads when they are compressed, and to distribute locomotor loads over a large area in order to keep foot tissue stresses within acceptable levels. In addition, the musculoskeletal foot arch and sole cushion of the elephant act in concert, similarly to the horse's cushioned frog and hoof and the human foot. In the elephant, the nearly half-cupula-shaped arrangement of the bony elements of the metatarsals and toes has interesting similarities to the structure of the arches of human feet.

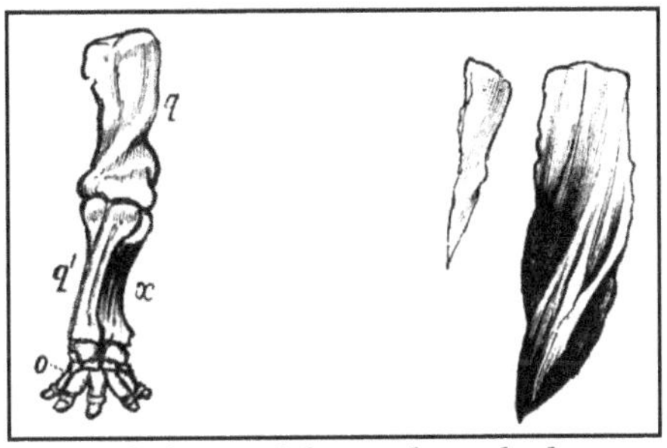

Leg skeleton of the modern elephant.

Recently, scientists at the Royal Veterinary College in the United Kingdom have discovered that the elephant possesses a sixth false toe, a sesamoid, located similarly to the giant panda's extra "thumb". They found that this sixth toe acts to support and distribute the weight of the elephant.

Human Foot

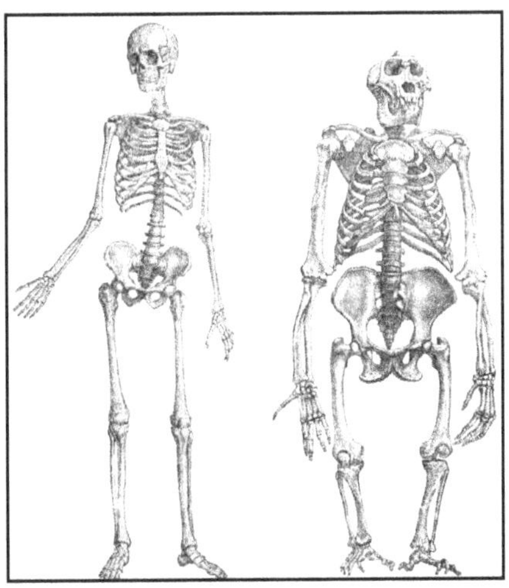
Skeleton of the human and gorilla (gorilla shown in non-natural posture).

The unique plantigrade alignment of the human foot results in a distal-limb structure that can adapt to a variety of conditions. The less mobile and more robust tarsal bones are shaped and aligned to accept and transmit large loads during the early phases of stance (initial contact and loading response phases of walking, and inadvertent heel strikes during running). The tarsals of the midfoot, which are smaller and shorter than the hindfoot tarsals, appear well oriented to transmit loads between the hindfoot and forefoot; this is necessary for load transfer and locking of the foot complex into a rigid lever for late stance phase. Conversely, the midfoot bones and joints also allow for the transmission of loads and inter-joint movement that unlocks the foot to create a loosely packed structure which renders the foot highly compliant over a variety of surfaces. In this configuration, the foot is able to absorb and damp the large loads encountered during heel strike and early weight acceptance. The forefoot, with its long metatarsal and relatively long phalanges, transmits loads during the end-of-stance phase that facilitate the push-off and transfer of forward momentum. The forefoot also serves as a lever to allow balance during standing and jumping. In addition, the arches of the foot that span the hindfoot, midfoot and forefoot play a critical role in the nature of transformation of the foot from a rigid lever to a flexible weight-accepting structure.

With a running gait, the foot-loading order is usually the reverse of walking. The foot strikes the ground with the ball of the foot, and then the heel drops. The heel drop elastically extends the Achilles tendon; this extension is reversed during the push-off.

Clinical Implications

Veterinarian or human healthcare professionals often respond when the foot of a dog, horse, elephant or human develops an abnormality. They typically investigate to understand the nature of the pathology in order to generate and implement a clinical treatment plan. For instance, the paws of the dog and the hindfoot work together to absorb the shock of jumping and running, and to provide flexibility of movement. If the dog's skeletal structures in areas other than the foot are compromised, the foot may be burdened with compensatory loading. Structural faults such as straight or loose shoulders, straight stifles, loose hips, and lack of balance between the forefoot and hindfoot, can all cause gait abnormalities that in turn damage the hindfoot and paws by overloading their foot structures as they compensate for the structural faults.

In the horse, dryness of the hoof may cause stiffening of the external foot structure. The stiffer hoof reduces the foot's load attenuation capacity, rendering the horse unable to bear much weight on the distal limb. Similar characteristic features emerge in the human foot in the form of the *pes cavus* alignment deformity, which is produced by tight connective tissue structures and joint congruency that create a rigid foot complex. Individuals with *pes cavus* display characteristic reduced load-attenuation features, and other structures proximal to the foot may compensate with increased load transfer (i.e., excessive loading to the knees, hips, lumbo-pelvic joints or lumbar vertebrae). Foot disorders are common in captive elephants. However, the cause is poorly understood.

Aquatic Locomotion

Aquatic locomotion is biologically propelled motion through a liquid medium. The simplest propulsive systems are composed of cilia and flagella. Swimming has evolved a number of times in a range of organisms including arthropods, fish, molluscs, reptiles, birds, and mammals.

Evolution of Swimming

Swimming evolved a number of times in unrelated lineages, and the evolutionary pressures leading to its adoption are unknown. Supposed jellyfish fossils occur in the Ediacaran, but the first free-swimming animals appear in the Early to Middle Cambrian. These are mostly related to the arthropods, and include the Anomalocaridids, which swam by means of lateral lobes in a fashion reminiscent of today's cuttlefish. Cephalopods joined the ranks of the nekton in the late Cambrian, and chordates were probably swimming from the Early Cambrian. Many terrestrial animals retain some capacity to swim, however some have returned to the water and developed the capacities for aquatic locomotion.

Micro-organisms

Bacterial

Ciliates

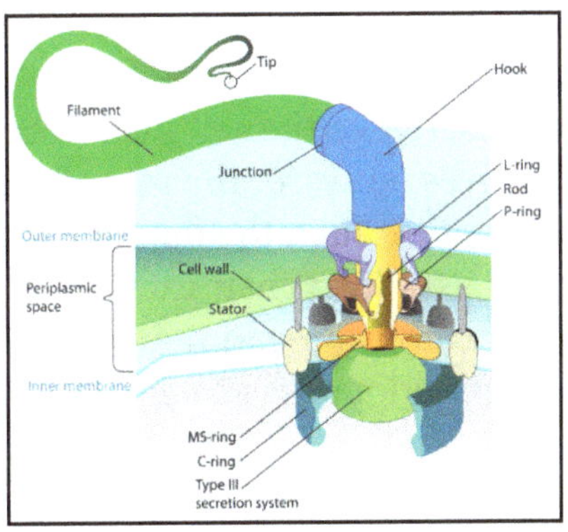

The flagellum of a Gram-negative bacteria is rotated by a molecular motor at its base.

Ciliates use small flagella called cilia to move through the water. One ciliate will generally have hundreds to thousands of cilia that are densely packed together in arrays. During movement, an individual cilium deforms using a high-friction power stroke followed by a low-friction recovery stroke. Since there are multiple cilia packed together on an individual organism, they display collective behavior in a metachronal rhythm. This means the deformation of one cilium is in phase with the deformation of its neighbor, causing deformation waves that propagate along the surface of the organism. These propagating waves of cilia are what allow the organism to use the cilia in a coordinated manner to move. A typical example of a ciliated microorganism is the *Paramecium*, a one-celled, ciliated protozoan covered by thousands of cilia. The cilia beating together allow the *Paramecium* to propel through the water at speeds of 500 micrometers per second.

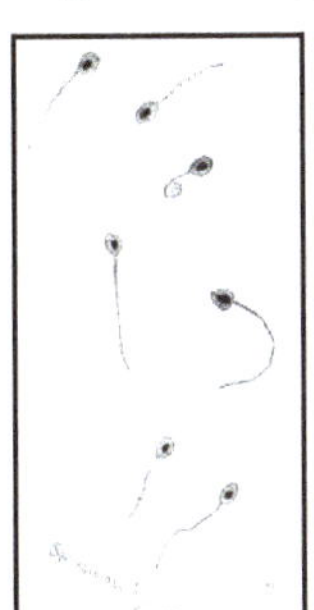

Salmon spermatozoa for artificial propagation.

Flagellates

Certain organisms such as bacteria and animal sperm have flagellum which have developed a way to move in liquid environments. A rotary motor model shows that bacteria uses the protons of an electrochemical gradient in order to move their flagella. Torque in the flagella of bacteria is created by particles that conduct protons around the base of the flagellum. The direction of rotation of the flagella in bacteria comes from the occupancy of the proton channels along the perimeter of the flagellar motor.

Movement of sperm is called sperm motility. The middle of the mammalian spermatozoon contains mitochondria that power the movement of the flagellum of the sperm. The motor around the base produces torque, just like in bacteria for movement through the aqueous environment.

Pseudopodia

Movement using a pseudopod is accomplished through increases in pressure at one point on the cell membrane. This pressure increase is the result of actin polymerization between the cortex and the membrane. As the pressure increases the cell membrane is pushed outward creating the pseudopod. When the pseudopod moves outward, the rest of the body is pulled forward by cortical tension. The result is cell movement through the fluid medium. Furthermore, the direction of movement is determined by chemotaxis. When chemoattraction occurs in a particular area of the cell membrane, actin polymerization can begin and move the cell in that direction. An excellent example of an organism that utilizes pseudopods is *Naegleria fowleri.*

A Simple Animation

Invertebrates

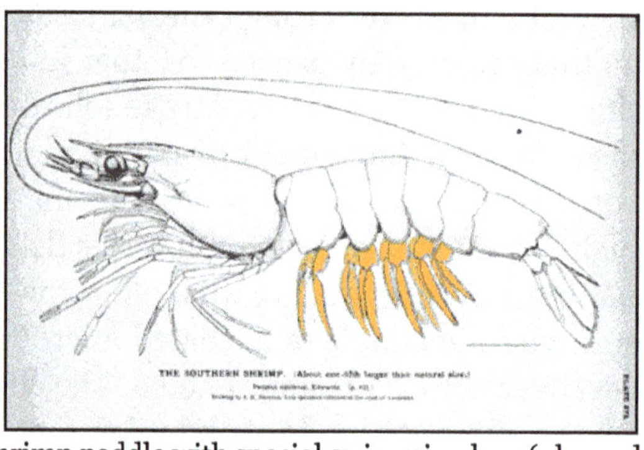

Shrimp paddle with special swimming legs (pleopods).

Among the radiata, jellyfish and their kin, the main form of swimming is to flex their cup shaped bodies. All jellyfish are free-swimming, although many of these spend most of their time swimming passively. Passive swimming is akin to gliding; the organism floats, using currents where it can, and does not exert any energy into controlling its position or motion. Active swimming, in contrast, involves the expenditure of energy to travel to a desired location.

In bilateria, there are many methods of swimming. The arrow worms (chaetognatha) undulate their finned bodies, not unlike fish. Nematodes swim by undulating their fin-less bodies. Some Arthropod groups can swim - including many crustaceans. Most crustaceans, such as shrimp, will usually swim by paddling with special swimming legs (pleopods). Swimming crabs swim with modified walking legs (pereiopods). Daphnia, a crustacean, swims by beating its antennae instead.

There are also a number of forms of swimming molluscs. Many free-swimming sea slugs, such as sea angels, flap fin-like structures. Some shelled molluscs, such as scallops can briefly swim by clapping their two shells open and closed. The molluscs most evolved for swimming are the cephalopods.

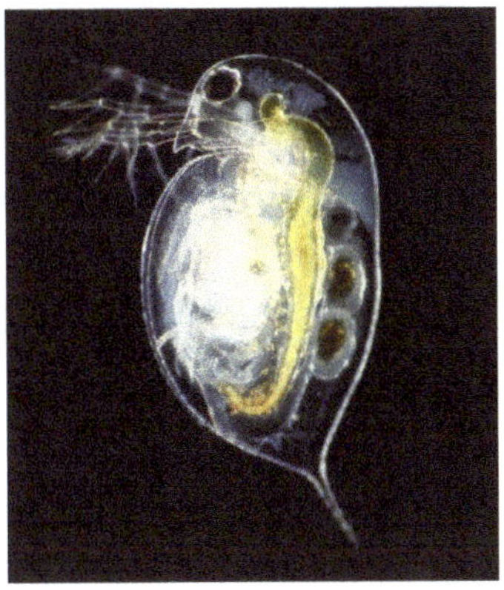

Daphnia swims by beating its antennae.

Among the Deuterostomia, there are a number of swimmers as well. Feather stars can swim by undulating their many arms . Salps move by pumping waters through their gelatinous bodies. The deuterosomes most evolved for swimming are found among the vertebrates, notably the fish.

Jet Propulsion

Octopuses swim headfirst, with arms trailing behind.

Jet propulsion is a method of aquatic locomotion where animals fill a muscular cavity and squirt out water to propel them in the opposite direction of the squirting water. Most organisms are equipped with one of two designs for jet propulsion; they can draw water from the rear and expel it from the rear, such as jellyfish, or draw water from front and expel it from the rear, such as salps. Filling up the cavity causes an increase in both the mass and drag of the animal. Because of the expanse of the contracting cavity, the animal's velocity fluctuates as it moves through the water, accelerating while expelling water and decelerating while vacuuming water. Even though these fluctuations in drag and mass can be ignored if the frequency of the jet-propulsion cycles is high enough, jet-propulsion is a relatively inefficient method of aquatic locomotion.

Jellyfish pulsate their bell for a type of jet locomotion.

Scallops swim by clapping their two shells open and closed.

All cephalopods can move by jet propulsion, but this is a very energy-consuming way to travel compared to the tail propulsion used by fish. The relative efficiency of jet propulsion decreases further as animal size increases. Since the Paleozoic, as competition with fish produced an environment where efficient motion was crucial to survival, jet propulsion has taken a back role, with fins and tentacles used to maintain a steady velocity. The stop-start motion provided by the jets, however, continues to be useful for providing bursts of high speed - not least when capturing prey or avoiding predators. Indeed, it makes cephalopods the fastest marine invertebrates, and they can outaccelerate most fish. Oxygenated water is taken into the mantle cavity to the gills and through muscular contraction of this cavity, the spent water is expelled through the hyponome, created by a fold in the mantle. Motion of the cephalopods is usually backward as water is forced out anteriorly through the hyponome, but direction can be controlled somewhat by pointing it in different directions. Most cephalopods float (i.e. are neutrally buoyant), so do not need to swim to remain afloat. Squid swim more slowly than fish, but use more power to generate their speed. The loss in efficiency is due to the amount of water the squid can accelerate out of its mantle cavity.

Jellyfish use a one-way water cavity design which generates a phase of continuous cycles of jet-propulsion followed by a rest phase. The Froude efficiency is about 0.09, which indicates a very costly method of locomotion. The metabolic cost of transport for jellyfish is high when compared to a fish of equal mass.

Other jet-propelled animals have similar problems in efficiency. Scallops, which use a similar design to jellyfish, swim by quickly opening and closing their shells, which draws in water and expels it from all sides. This locomotion is used as a means to escape predators such as starfish. Afterwards, the shell acts as a hydrofoil to counteract the scallop's tendency to sink. The Froude efficiency is low for this type of movement, about 0.3, which is why it's used as an emergency escape mechanism from predators. However, the amount of work the scallop has to do is mitigated by the elastic hinge that connects the two shells of the bivalve. Squids swim by drawing water into their mantle cavity and expelling it through their siphon. The Froude efficiency of their jet-propulsion system is around 0.29, which is much lower than a fish of the same mass.

Much of the work done by scallop muscles to close its shell is stored as elastic energy in abductin tissue, which acts as a spring to open the shell. The elasticity causes the work done against the water to be low because of the large openings the water has to enter and the small openings the water has to leave. The inertial work of scallop jet-propulsion is also low. Because of the low inertial work, the energy savings created by the elastic tissue is so small that it's negligible. Medusae can also use their elastic mesoglea to enlarge their bell. Their mantle contains a layer of muscle sandwiched between elastic fibers. The muscle fibers run around the bell circumferentially while the elastic fibers run through the muscle and along the sides of the bell to prevent lengthening. After making a single contraction, the bell vibrates passively at the resonant frequency to refill the bell. However, in contrast with scallops, the inertial work is similar to the hydrodynamic work due to how medusas expel water - through a large opening at low velocity. Because of this, the negative pressure created by the vibrating cavity is lower than the positive pressure of the jet, meaning that inertial work of the mantle is small. Thus, jet-propulsion is shown as an inefficient swimming technique.

Fish

Open water fish, like this Atlantic bluefin tuna, are usually streamlined for straightline speed, with a deeply forked tail and a smooth body shaped like a spindle tapered at both ends.

Many reef fish, like this queen angelfish, have a body flattened like a pancake, with pectoral and pelvic fins that act with the flattened body to maximize manoeuvrability.

Many fish swim through water by creating undulations with their bodies or oscillating their fins. The undulations create components of forward thrust complemented by a rearward force, side forces which are wasted portions of energy, and a normal force that is between the forward thrust and side force. Different fish swim by undulating different parts of their bodies. Eel-shaped fish undulate their entire body in rhythmic sequences. Streamlined fish, such as salmon, undulate the caudal portions of their bodies. Some fish, such as sharks, use stiff, strong fins to create dynamic lift and propel themselves. It is common for fish to use more than one form of propulsion, although they will display one dominant mode of swimming Gait changes have even been observed in juvenile reef fish of various sizes. Depending on their needs, fish can rapidly alternate between synchronized fin beats and alternating fin beats.

According to *Guinness World Records 2009*, *H. zosterae* (the dwarf seahorse) is the slowest moving fish, with a top speed of about 5 feet (150 cm) per hour. They swim very poorly, rapidly fluttering a dorsal fin and using pectoral fins (located behind their eyes) to steer. Seahorses have no caudal fin.

Body-caudal Fin (BCF) Propulsion

- Anguilliform: Anguilliform swimmers are typically slow swimmers. They undulate the majority of their body and use their head as the fulcrum for the load they are moving. At any point during their undulation, their body has an amplitude between 0.5-1.0 wavelengths. The amplitude that they move their body through allows them to swim backwards. Anguilliform locomotion is usually seen in fish with long, slender bodies like eels, lampreys, oarfish, and a number of catfish species.
- Subcarangiform, Carangiform, Thunniform: These swimmers undulate the posterior half of their body and are much faster than anguilliform swimmers. At any point while they are swimming, a wavelength <1 can be seen in the undulation pattern of the body. Some Carangiform swimmers include nurse sharks, bamboo sharks, and reef sharks. Thunniform swimmers are very fast and some common Thunniform swimmers include tuna, white sharks, salmon, jacks, and mako sharks. Thunniform swimmers only undulate their high aspect ratio caudal fin, so they are usually very stiff to push more water out of the way.
- Ostraciiform: Ostraciiform swimmers oscillate their caudal region, making them relatively slow swimmers. Boxfish, torpedo rays, and momyrs employ Ostraciiform locomotion. The cow fish uses Osctraciiform locomotion to hover in the water column.

Median Paired Fin (MPF) Propulsion

- Tetraodoniform, Balistiform, Diodontiform: These swimmers oscillate their median fins. They are typically slow swimmers, and some notable examples include the oceanic sunfish (which has extremely modified anal and dorsal fins), puffer fish, and triggerfish.
- Rajiform, Amiiform, Gymnotiform: This locomotory mode is accomplished by undulation of the pectoral and median fins. During their undulation pattern, a wavelength >1 can be seen in their fins. They are typically slow to moderate swimmers, and some examples include rays, bowfin, and knife fishes. The black ghost knife fish is a Gymnotiform swimmer that has a very long ventral ribbon fin. Thrust is produced by passing waves down the ribbon fin while the body remains rigid. This also allows the ghost knife fish to swim in reverse.
- Labriform: Labriform swimmers are also slow swimmers. They oscillate their pectoral fins to create thrust. Oscillating fins create thrust when a starting vortex is shed from the trailing edge of the fin. As the foil departs from the starting vortex, the effect of that vortex diminishes, while the bound circulation remains, producing lift. Labriform swimming can be viewed as continuously starting and stopping. Wrasses and surf perch are common Labriform swimmers.

Hydrofoils

The leopard shark angles its pectoral fins so they behave as hydrofoils to control the animal's pitch.

Hydrofoils, or fins, are used to push against the water to create a normal force to provide thrust, propelling the animal through water. Sea turtles and penguins beat their paired hydrofoils to create lift. Some paired fins, such as pectoral fins on leopard sharks, can be angled at varying degrees to allow the animal to rise, fall, or maintain its level in the water column. The reduction of fin surface area helps to minimize drag, and therefore increase efficiency. Regardless of size of the animal, at any particular speed, maximum possible lift is proportional to (wing area) x $(\text{speed})^2$. Dolphins and whales have large, horizontal caudal hydrofoils, while many fish and sharks have vertical caudal hydrofoils. Porpoising (seen in cetaceans, penguins, and pinnipeds) may save energy if they are moving fast. Since drag increases with speed, the work required to swim unit distance is greater at higher speeds, but the work needed to jump unit distance is independent of speed. Seals propel themselves through the water with their caudal tail, while sea lions create thrust solely with their pectoral flippers.

Drag Powered Swimming

The slowest-moving fishes are the sea horses, often found in reefs.

As with moving through any fluid, friction is created when molecules of the fluid collide with organism. The collision causes drag against moving fish, which is why many fish are streamlined in shape. Streamlined shapes work to reduce drag by orienting elongated objects parallel to the force of drag, therefore allowing the current to pass over and taper off the end of the fish. This streamlined shape allows for more efficient use of energy locomotion. Some flat-shaped fish can take advantage of pressure drag by having a flat bottom surface and curved top surface. The pressure drag created allows for the upward lift of the fish.

Appendages of aquatic organisms propel them in two main and biomechanically extreme mechanisms. Some use lift powered swimming, which can be compared to flying as appendages flap like wings, and reduce drag on the surface of the appendage. Others use drag powered swimming, which can be compared to oars rowing a boat, with movement in a horizontal plane, or paddling, with movement in the parasagittal plane.

Drag swimmers use a cyclic motion where they push water back in a power stroke, and return their limb forward in the return or recovery stroke. When they push water directly backwards, this moves their body forward, but as they return their limbs to the starting position, they push water forward, which will thus pull them back to some degree, and so opposes the direction that the body is heading. This opposing force is called drag. The return-stroke drag causes drag swimmers to employ different strategies than lift swimmers. Reducing drag on the return stroke is essential for optimizing efficiency. For example, ducks paddle through the water spreading the webs of their feet as they move water back, and then when they return their feet to the front they pull their webs together to reduce the subsequent pull of water forward. The legs of water beetles have little hairs which spread out to catch up and move water back in the power stroke, but lay flat as the appendage moves forward in the return stroke. Also, the water beetle's legs have a side that is wider and is held perpendicular to the motion when pushing backward, but the leg is then rotated when the limb is to return forward, so that the thinner side will catch up less water.

Drag swimmers experience a lessened efficiency in swimming due to resistance which affects their optimum speed. The less drag a fish experiences, the more it will be able to maintain higher speeds. Morphology of the fish can be designed to reduce drag, such as streamlining the body. The cost of transport is much higher for the drag swimmer, and when deviating from its optimum speed, the drag swimmer is energetically strained much more than the lift swimmer. There are natural processes in place to optimize energy use, and it is thought that adjustments of metabolic rates can compensate in part for mechanical disadvantages.

Semi-aquatic animals compared to fully aquatic animals exhibit exacerbation of drag. Design that allows them to function out of the water limits the efficiency possible to be reached when in the water. In water swimming at the surface exposes them to resistive wave drag and is associated with a higher cost than submerged swimming. Swimming below the surface exposes them to resistance due to return strokes and pressure, but primarily friction. Frictional drag is due to fluid viscosity and morphology characteristics. Pressure drag is due to the difference of water flow around the body and is also affected by body morphology. Semi-aquatic organisms encounter increased resistive forces when in or out of the water, as they are not specialized for either habitat. The morphology of otters and beavers, for example, must meet needs for both environments. Their fur decreases streamlining and creates additional drag. The platypus may be a good example of an intermediate between drag and lift swimmers because it has been shown to have a rowing mechanism which is similar to lift-based pectoral oscillation. The limbs of semi-aquatic organisms are reserved for use on land and using them in water not only increases the cost of locomotion, but limits them to drag-based modes. Although they are less efficient, drag swimmers are able to produce more thrust at low speeds than lift swimmers. They are also thought to be better for maneuverability due to the large thrust produced.

Amphibians

Most of the Amphibia have a larval state, which has inherited anguilliform motion, and a laterally compressed tail to go with it, from fish ancestors. The corresponding tetrapod adult forms, even in the tail-retaining sub-class Urodeles, are sometimes aquatic to only a negligible extent (as in the genus Salamandra, whose tail has lost its suitability for aquatic propulsion), but the majority of Urodeles, from the newts to the giant salamander Megalobatrachus, retain a laterally compressed tail for a life that is aquatic to a considerable degree, which can use in a carangiform motion.

Common toad (Bufo bufo) swimming.

Of the tailless amphibians (the frogs and toads of the sub-class Anura) the majority are aquatic to an insignificant extent in adult life, but in that considerable minority that are mainly aquatic we encounter for the first time the problem of adapting the tailless-tetrapod structure for aquatic propulsion. The mode that they use is unrelated to any used by fish. With their flexible back legs and webbed feet they execute something close to the leg movements of a human 'breast stroke,' rather more efficiently because the legs are better streamlined.

Reptiles

Nile crocodile (Crocodylus niloticus) swimming.

From the point of view of aquatic propulsion, the descent of modern members of the class Reptilia from archaic tailed Amphibia is most obvious in the case of the order Crocodilia (crocodiles and alligators), which use their deep, laterally compressed tails in an essentially carangiform mode of propulsion.

Terrestrial snakes, in spite of their 'bad' hydromechanical shape with roughly circular cross-section and gradual posterior taper, swim fairly readily when required, by an anguilliform propulsion.

Cheloniidae (true turtles) have found a beautiful solution to the problem of tetrapod swimming through the development of their forelimbs into flippers of high-aspect-ratio wing shape, with which they imitate a bird's propulsive mode more accurately than do the eagle-rays themselves.

Immature Hawaiian green sea turtle in shallow waters.

Macroplata.

Fin and Flipper Locomotion

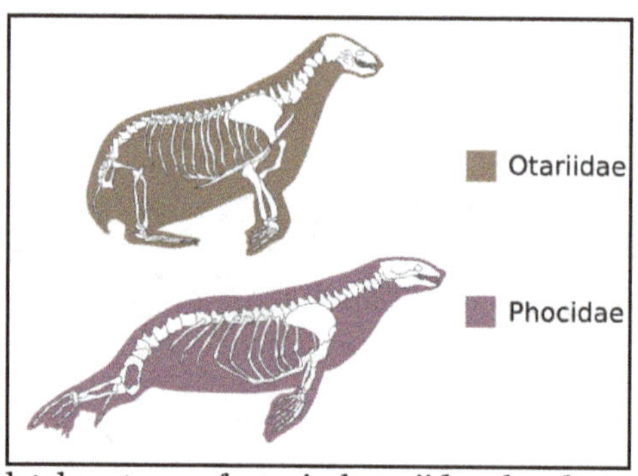

Comparative skeletal anatomy of a typical otariid seal and a typical phocid seal.

Aquatic reptiles such as sea turtles and extinct species like Pliosauroids predominantly use their pectoral flippers to propel themselves through the water and their pelvic flippers for maneuvering. During swimming they move their pectoral flippers in a dorso-ventral motion, causing forward motion. During swimming, they rotate their front flippers to decrease drag through the water column and increase efficiency. Newly hatched sea turtles exhibit several behavioral skills that help orientate themselves towards the ocean as well as identifying the transition from sand to water. If rotated in the pitch, yaw or roll direction, the hatchlings are capable of counteracting the forces acting upon them by correcting with either their pectoral or pelvic flippers and redirecting themselves towards the open ocean.

Among mammals otariids (fur seals) swim primarily with their front flippers, using the rear flippers for steering, and phocids (true seals) move the rear flippers laterally, pushing the animal through the water.

Escape Reactions

Animated representation of lobstering.

Some arthropods, such as lobsters and shrimps, can propel themselves backwards quickly by flicking their tail, known as lobstering or the caridoid escape reaction.

Varieties of fish, such as teleosts, also use fast-starts to escape from predators. Fast-starts are characterized by the muscle contraction on one side of the fish twisting the fish into a C-shape. Afterwards, muscle contraction occurs on the opposite side to allow the fish to enter into a steady swimming state with waves of undulation traveling alongside the body. The power of the bending motion comes from fast-twitch muscle fibers located in the central region of the fish. The signal to perform this contraction comes from a set of Mauthner cells which simultaneously send a signal to the muscles on one side of the fish. Mauthner cells are activated when something startles the fish and can be activated by visual or sound-based stimuli.

Fast-starts are split up into three stages. Stage one, which is called the preparatory stroke, is characterized by the initial bending to a C-shape with small delay caused by hydrodynamic resistance. Stage two, the propulsive stroke, involves the body bending rapidly to the other side, which may occur multiple times. Stage three, the rest phase, cause the fish to return to normal steady-state swimming and the body undulations begin to cease. Large muscles located closer to the central portion of the fish are stronger and generate more force than the muscles in the tail. This asymmetry in muscle composition causes body undulations that occur in Stage 3. Once the fast-start is completed, the position of the fish has been shown to have a certain level of unpredictability, which helps fish survive against predators.

The rate at which the body can bend is limited by resistance contained in the inertia of each body part. However, this inertia assists the fish in creating propulsion as a result of the momentum created against the water. The forward propulsion created from C-starts, and steady-state swimming in general, is a result of the body of the fish pushing against the water. Waves of undulation create rearward momentum against the water providing the forward thrust required to push the fish forward.

Efficiency

The Froude propulsion efficiency is defined as the ratio of power output to the power input:

$$nf = 2U_1 / (U_1 + U_2)$$

where U1 = free stream velocity and U2 = jet velocity. A good efficiency for carangiform propulsion is between 50 and 80%.

Minimizing Drag

Pressure differences occur outside the boundary layer of swimming organisms due to disrupted flow around the body. The difference on the up- and down-stream surfaces of the body is pressure drag, which creates a downstream force on the object. Frictional drag, on the other hand, is a result of fluid viscosity in the boundary layer. Higher turbulence causes greater frictional drag.

Reynolds number (Re) is the measure of the relationships between inertial and viscous forces in flow ((animal's length x animal's velocity)/kinematic viscosity of the fluid). Turbulent flow can be found at higher Re values, where the boundary layer separates and creates a wake, and laminar flow can be found at lower Re values, when the boundary layer separation is delayed, reducing wake and kinetic energy loss to opposing water momentum.

The body shape of a swimming organism affects the resulting drag. Long, slender bodies reduce pressure drag by streamlining, while short, round bodies reduce frictional drag; therefore, the optimal shape of an organism depends on its niche. Swimming organisms with a fusiform shape are likely to experience the greatest reduction in both pressure and frictional drag.

Wing shape also affects the amount of drag experienced by an organism, as with different methods of stroke, recovery of the pre-stroke position results in the accumulation of drag.

High-speed ram ventilation creates laminar flow of water from the gills along the body of an organism.

The secretion of mucus along the organism's body surface, or the addition of long-chained polymers to the velocity gradient, can reduce frictional drag experienced by the organism.

Buoyancy

Many aquatic/marine organisms have developed organs to compensate for their weight and control their buoyancy in the water. These structures, make the density of their bodies very close to that of the surrounding water. Some hydrozoans, such as siphonophores, has gas-filled floats; the Nautilus, Sepia, and Spirula (Cephalopods) have chambers of gas within their shells; and most teleost fish and many lantern fish (Myctophidae) are equipped with swim bladders. Many aquatic and marine organisms may also be composed of low-density materials. Deep-water teleosts, which do not have a swim bladder, have few lipids and proteins, deeply ossified bones, and watery tissues that maintain their buoyancy. Some sharks' livers are composed of low-density lipids, such as hydrocarbon squalene or wax esters (also found in Myctophidae without swim bladders), which provide buoyancy.

Swimming animals that are denser than water must generate lift or adapt a benthic lifestyle. Movement of the fish to generate hydrodynamic lift is necessary to prevent sinking. Often, their bodies act as hydrofoils, a task that is more effective in flat-bodied fish. At a small tilt angle, the lift is greater for flat fish than it is for fish with narrow bodies. Narrow-bodied fish use their fins as hydrofoils while their bodies remain horizontal. In sharks, the heterocercal tail shape drives water downward, creating a counteracting upward force while thrusting the shark forward. The lift generated is assisted by the pectoral fins and upward-angle body positioning. It is supposed that tunas primarily use their pectoral fins for lift.

Buoyancy maintenance is metabolically expensive. Growing and sustaining a buoyancy organ, adjusting the composition of biological makeup, and exerting physical strain to stay in motion demands large amounts of energy. It is proposed that lift may be physically generated at a lower energy cost by swimming upward and gliding downward, in a "climb and glide" motion, rather than constant swimming on a plane.

Temperature

Temperature can also greatly affect the ability of aquatic organisms to move through water. This is because temperature not only affects the properties of the water, but also the organisms in the water, as most have an ideal range specific to their body and metabolic needs.

Q10 (temperature coefficient), the factor by which a rate increases at a 10 °C increase in temperature, is used to measure how organisms' performance relies on temperature. Most have increased rates as water becomes warmer, but some have limits to this and others find ways to alter such effects, such as by endothermy or earlier recruitment of faster muscle.

For example, *Crocodylus porosus*, or estuarine crocodiles, were found to increase swimming speed from 15 °C to 23 °C and then to have peak swimming speed from 23 °C to 33 °C. However, performance began to decline as temperature rose beyond that point, showing a limit to the range of temperatures at which this species could ideally perform.

Submergence

The more of the animal's body that is submerged while swimming, the less energy it uses. Swimming on the surface requires two to three times more energy than when completely submerged. This is because of the bow wave that is formed at the front when the animal is pushing the surface of the water when swimming, creating extra drag.

Secondary Evolution

Chinstrap penguin leaping over water.

While tetrapods lost many of their natural adaptations to swimming when they evolved onto the land, many have re-evolved the ability to swim or have indeed returned to a completely aquatic lifestyle.

Primarily or exclusively aquatic animals have re-evolved from terrestrial tetrapods multiple times: examples include amphibians such as newts, reptiles such as crocodiles, sea turtles, ichthyosaurs, plesiosaurs and mosasaurs, marine mammals such as whales, seals and otters, and birds such as penguins. Many species of snakes are also aquatic and live their entire lives in the water. Among invertebrates, a number of insect species have adaptations for aquatic life and locomotion. Examples of aquatic insects include dragonfly larvae, water boatmen, and diving beetles. There are also aquatic spiders, although they tend to prefer other modes of locomotion under water than swimming proper.

Swimming dog

Even though primarily terrestrial tetrapods have lost many of their adaptations to swimming, the ability to swim has been preserved or re-developed in many of them. It may never have been completely lost.

Examples are: Some breeds of dog swim recreationally. Umbra, a world record-holding dog, can swim 4 miles (6.4 km) in 73 minutes, placing her in the top 25% in human long-distance swimming competitions. Although most cats hate water, adult cats are good swimmers. The fishing cat is one wild species of cat that has evolved special adaptations for an aquatic or semi-aquatic lifestyle – webbed digits. Tigers and some individual jaguars are the only big cats known to go into water readily, though other big cats, including lions, have been observed swimming. A few domestic cat breeds also like swimming, such as the Turkish Van.

Horses, moose, and elk are very powerful swimmers, and can travel long distances in the water. Elephants are also capable of swimming, even in deep waters. Eyewitnesses have confirmed that camels, including dromedary and Bactrian camels, can swim, despite the fact that there is little deep water in their natural habitats.

Both domestic and wild rabbits can swim. Domestic rabbits are sometimes trained to swim as a circus attraction. A wild rabbit famously swam in an apparent attack on U.S. President Jimmy Carter's boat when it was threatened in its natural habitat.

The guinea pig (or cavy) is noted as having an excellent swimming ability. Mice can swim quite well. They do panic when placed in water, but many lab mice are used in the Morris water maze, a test to measure learning. When mice swim, they use their tails like flagella and kick with their legs.

Many snakes are excellent swimmers as well. Large adult anacondas spend the majority of their time in the water, and have difficulty moving on land.

Many monkeys can naturally swim and some, like the proboscis monkey, crab-eating macaque, and rhesus macaque swim regularly.

Human Swimming

Swimming has been known amongst humans since prehistoric times; the earliest record of swimming dates back to Stone Age paintings from around 7,000 years ago. Competitive swimming started in Europe around 1800 and was part of the first modern 1896 Summer Olympics in Athens, though not in a form comparable to the contemporary events. It was not until 1908 that regulations were implemented by the International Swimming Federation to produce competitive swimming.

Fin and Flipper Locomotion

Periophthalmus gracilis (from Malaysia to North Australia).

Fin and flipper locomotion occurs mostly in aquatic locomotion, and rarely in terrestrial locomotion. From the three common states of matter — gas, liquid and solid, these appendages are adapted for liquids, mostly Fresh or Saltwater and used in locomotion, steering and balancing of the body. Locomotion is important in order to escape predators, acquire food, find mates and bury for shelter, nest or food. Aquatic locomotion consists of swimming, whereas terrestrial locomotion encompasses walking, 'crutching', jumping, digging as well as covering. Some animals such as Sea turtles and Mudskippers use these two environments for different purposes, for example using the land for nesting, and the sea to hunt for food.

Aquatic Locomotion with Fins and Flippers

Aquatic Locomotion of Fish

Fish live in Fresh or Saltwater habitats and some exceptions are capable of coming on land (Mudskippers). Most fish have a line of muscle blocks, called myomeres, along each side of the body. To swim, they alternately contract one side and relax the other side in a progression which goes from the head to the tail. In this way, an undulatory locomotion results, first bending the body one way in a wave which travels down the body, and then back the other way, with the contracting and relaxing muscles switching roles. They use their fins to propel themselves through the water in this swimming motion. Actinopterygians, the ray-finned fish show an evolutionary pattern of fine control ability to control the dorsal and ventral lobe of the caudal fin. Through developmental changes, intrinsic caudal muscles were added, which enable fish to exhibit such complex maneuvers such as control during acceleration, braking and backing. Studies have shown that the muscles in the caudal fin, have independent activity patterns from the myotomal musculature. These results show specific kinematic roles for different part of the fish's musculature. A curious example of fish adaption is the Ocean sunfish, also known as the *Mola mola*. These fish have undergone significant developmental changes reducing their spinal cord, giving them a disk like appearance, and investing in two very large fins for propulsion. This adaptation usually gives them the appearance that they are as long as they are tall. They are also amazing fish in that they hold the world record in weight gain from fry to adult (60 million times its weight).

Aquatic Locomotion of Marine Mammals

Swimming mammals, such as whales, dolphins,and sea lions, use their flippers to move forward through the water column. During swimming sea lions have a thrust phase, which lasts about 60% of the full cycle, and the recovery phase lasts the remaining 40%. A full cycle duration lasts about 0.5 to 1.0 seconds. Changing direction is a very rapid maneuver that is initiated by head movement towards the back of the animal that is followed by a spiral turn with the body. Due to their pectoral flippers being so closely located to their center of gravity, sea lions are capable of displaying astounding maneuverability in the pitch, roll, and yaw direction and are therefore not constrained, turning stochastically as they please. It is hypothesized that the increased level of maneuverability is caused by their complex habitat. Hunting occurs in difficult environments containing rocky inshore/kelp forest communities, with many niches for prey to hide, therefore requiring speed and maneuverability for capture. The complex skills of a sea lion are learned early on in ontogeny and most are perfected by the time the pups reach one year. Whales and dolphins are less maneuverable and more constrained in their movements. However, dolphins are capable of accelerating as fast as sea lions, but they are not capable of turning as quickly and as efficiently. For both whales and dolphins, their center of gravity does not line up with their pectoral flippers in a straight line, causing a much more rigid and stable swimming pattern.

Aquatic Locomotion of Marine Reptiles

Aquatic reptiles such as sea turtles predominantly use their pectoral flippers to propulse through the water and their pelvic flippers for maneuvering. During swimming they move their pectoral flippers in a clapping motion underneath their body and pull them back up into an airplane position, causing forward motion. During the swimming motion it is really important that they rotate their front flipper in order to decrease drag through the water column and increase their efficiency. Sea turtles exhibit a natural suite of behavior skills that help them direct themselves towards the ocean as well as identify the transition from sand to water after hatching. If rotated in the pitch, yaw or roll direction the hatchlings are capable of counteracting the forces acting upon them by correcting with either their pectoral or pelvic flippers and redirecting themselves towards the open ocean.

Terrestrial Locomotion

Terrestrial Locomotion of Fish

Terrestrial locomotion poses new obstacles such as gravity and new media, including sand, mudd, twigs, logs, debris, grass and many more. Fins and flippers are aquatically adapted appendages and typically aren't very useful in such an environment. It could be hypothesized that fish would try to "swim" on land, but studies have shown that some fish evolved to cope with the terrestrial environment. Mudskippers, for example demonstrate a 'crutching' gait which enables them to 'walk' over muddy surfaces as well as dig burrows to hide in. Mudskippers are also able to jump up to 3 cm distances. This behavior is described as starting with a J-curvature of the body at about

2/3 of its body length (with its tail wrapped towards the head), followed by a straightening of their body which propulses them like a projectile through the air. This behavior enables them to cope with the new environment and opens their habitat to new food sources as well as new predators.

Mudskippers in The Gambia.

Terrestrial Locomotion of Marine Reptiles

Caretta caretta Jekyll Island, GA.

Reptiles, such as sea turtles spend most of their lives in the ocean. However, their life cycle requires the females to come on shore and lay their nests on the beach. Consequently, the hatchlings emerge from the sand and have to run toward the water. Depending on their species, sea turtles are described to have either a symmetrical gait (diagonally opposite limbs are moving together) or an asymmetrical gait (Contra-lateral limbs move together). For example, loggerhead sea turtle hatchlings are commonly seen exhibiting symmetrical gait on sand, whereas, leatherback sea turtles employ the asymmetrical gait while on land. Notably, leatherbacks employ their front (pelvic) flippers more during forward terrestrial locomotion. Sea turtles can be seen nesting on subtropical and tropical beaches all around the world and exhibit such behavior such as arribada (Collective animal behavior). This is a phenomenon seen in Kemp's Ridley turtles which emerge all at once in one night only onto the beach to lay their nests.

References

- Jurmain, Robert; Kilgore, Lynn; Trevathan, Wenda (2008). Essentials of Physical Anthropology (7 ed.). Cengage Learning. p. 109. ISBN 9780495509394
- Davies, S.J.J.F. (2003). "Birds I Tinamous and Ratites to Hoatzins". In Hutchins, Michael. Grzimek's Animal Life Encyclopedia. 8 (2 ed.). Farmington Hills, MI: Gale Group. pp. 99–101. ISBN 0-7876-5784-0

- Dorit, R. L.; Walker, W. F.; Barnes, R. D. (1991). Zoology. Saunders College Publishing. p. 782. ISBN 978-0-03-030504-7
- King, R.S. (2013). "BiLBIQ: A Biologically Inspired Robot with Walking and Rolling Locomotion". Biosystems and Biorobotics. 2. Springer, Verlag, Berlin, Heidelberg. doi:10.1007/978-3-642-34682-8. ISBN 978-3-642-34681-1
- Harkness, John E.; Wagner, Joseph E. (1995). The Biology and Medicine of Rabbits and Rodents. Williams & Wilkins. pp. 30–39. ISBN 0-683-03919-9
- Kehm, G (2007). Olympic Swimming and Diving:Great Moments in Olympic History. The Rosen Publishing Group. pp. 14. ISBN 1-4042-0970-0
- Daniel L. Gilbert; William J. Adelman; John M. Arnold (1990). Squid as Experimental Animals. Daniel L. Gilbert, John M. Arnold (illustrated ed.). Springer,. ISBN 978-0-306-43513-3
- Weis, J.S. (2012). Walking sideways: the remarkable world of crabs. Ithaca, NY: Cornell University Press. pp. 63–77. ISBN 978-0-8014-5050-1. OCLC 794640315
- Campbell, Neil A.; Reece, Jane B. (2005). Biology, 7th Edition. San Francisco: Pearson - Benjamin Cummings. pp. 522–523. ISBN 0-8053-7171-0

Permissions

All chapters in this book are published with permission under the Creative Commons Attribution Share Alike License or equivalent. Every chapter published in this book has been scrutinized by our experts. Their significance has been extensively debated. The topics covered herein carry significant information for a comprehensive understanding. They may even be implemented as practical applications or may be referred to as a beginning point for further studies.

We would like to thank the editorial team for lending their expertise to make the book truly unique. They have played a crucial role in the development of this book. Without their invaluable contributions this book wouldn't have been possible. They have made vital efforts to compile up to date information on the varied aspects of this subject to make this book a valuable addition to the collection of many professionals and students.

This book was conceptualized with the vision of imparting up-to-date and integrated information in this field. To ensure the same, a matchless editorial board was set up. Every individual on the board went through rigorous rounds of assessment to prove their worth. After which they invested a large part of their time researching and compiling the most relevant data for our readers.

The editorial board has been involved in producing this book since its inception. They have spent rigorous hours researching and exploring the diverse topics which have resulted in the successful publishing of this book. They have passed on their knowledge of decades through this book. To expedite this challenging task, the publisher supported the team at every step. A small team of assistant editors was also appointed to further simplify the editing procedure and attain best results for the readers.

Apart from the editorial board, the designing team has also invested a significant amount of their time in understanding the subject and creating the most relevant covers. They scrutinized every image to scout for the most suitable representation of the subject and create an appropriate cover for the book.

The publishing team has been an ardent support to the editorial, designing and production team. Their endless efforts to recruit the best for this project, has resulted in the accomplishment of this book. They are a veteran in the field of academics and their pool of knowledge is as vast as their experience in printing. Their expertise and guidance has proved useful at every step. Their uncompromising quality standards have made this book an exceptional effort. Their encouragement from time to time has been an inspiration for everyone.

The publisher and the editorial board hope that this book will prove to be a valuable piece of knowledge for students, practitioners and scholars across the globe.

Index

www.ingramcontent.com/pod-product-compliance
Lightning Source LLC
LaVergne TN
LVHW080911210926
848485LV00004B/628

* 9 7 8 1 6 3 9 8 9 3 0 1 0 *